EQUATION

Ulas Can Cengiz

CHAPTER ONE

The Idea Has Arrived

What to do Before Planning Your Next App

Think about the situation you're in right now. Where do you live? What are your resources? Can you afford to pay salaries? How long can you support your family? These are the questions you need to answer before planning an app. Because it's not only about executing your plans, but also about how you achieve them.

The most critical thing may be the execution, but your situation determines your execution capability. For example, if you have sufficient capital you don't need to handle everything by yourself. Or if you don't have the available time to set up a startup, then first you'll have to look at managing your time.

Know your exact situation right from the start. That's what they don't tell you in success stories. You're different from anyone else. Everyone is. So, you have to be familiar with your unique status, and then, and only then, is the time to start a new project, even if it's only a side project.

Knowing what you have and what you lack is important, and it's probably more important than you think. Because nothing just happens when you begin a startup, you are responsible for making things happen. Without your energy, there will be nothing.

Also, no one can execute your idea like you do. But it's not only about having the idea, it's about having the details and you need to fill in the gaps between each and every detail. To do that, you need to know what's going on around you.

Let me give you an example: Two people I know had a remarkably similar idea about creating a social network for students. Both of them were university students and both of them had enough money to build the product.

They weren't developers, they weren't designers. They had a little knowledge about those topics, but that's all. Then they started to think about their idea.

The first one, let's call him Mike, formed a team. He did the branding, went to university clubs to test the idea, communicated it to as many people as he could, and tried to spread the word throughout his circle.

The second, let's call him Jonathan, took a different approach. He briefed a development team so they could start building it. Then he began to create a brand with a small agency.

When Jonathan's app had been developed, Mike's app had not yet got off the ground. Jonathan made a simple discovery that day: No one in the world can read your mind so you must tell them what you want.

Mike's app was developed four months after Jonathan's. It was a brief success. It was not as successful as you might expect, but it was way better than Jonathan's app. Basically, that showed me the importance of the details.

The more details you have, and the more you communicate them to the people around you, matters. Another important aspect is the aptitude of those people with whom you share your idea. What are their capabilities? How can they make your idea shine?

That's why you shouldn't ask what your family or close friends think about your idea. The point is, you're important to these people. They wouldn't want you to be demotivated so they can't be brutally honest.

* * *

You need genuine feedback from honest, talented people who are unburdened by any emotional connection to you. In this way, you can generate a constructive response to your ideas and so improve yourself and your thinking.

You may not have someone around you to get feedback from; then, it's a sign that you should make more connections. If you want to be successful as an entrepreneur, you need people to give you money.

And the people whose money you want will be thinking about your product, not about you. You should have established a way to approach them.

We will come to the important design and the development phases later, however, knowing yourself and your capabilities are more important than any of those.

For example, if you can't manage people, either you must learn how to, or you will need to employ a manager or a consultant to do it for you. Or, if you trust yourself on visuals, you need to battle-test that ability of yours under trying conditions.

I can't emphasize enough how important it is for you to know your own strengths and weaknesses, but there is one more thing: influencing and convincing people is a hard thing to do, and you will have to master it if you want to build a new product successfully, without having to do everything yourself.

Because it's not like paying a team to develop something magical, there is no magic in development, it's just code that people write, keystroke by keystroke. I have developed many apps, and I assure you there is no magic in the process at all.

The developers will need information, not only about your idea but also about the details. We'll take a look at how you obtain more information about the details in the upcoming chapters, so keep reading.

Like your team, you'll need information too, and if you haven't already, you should learn to ask for it. Because, believe me, you don't know that much about your idea right now. You need to learn about the conditions in which you can execute it.

Let's shine some light on the process and make it clearer:

Say you are a yoga trainer, and you have an idea to build a yoga app. You want it to be the best in the market, and you want to make it quickly. The first thing you need to do is thinking about the challenges your clients face every day.

Not just think about them, ask them. Ask them about their dream yoga app, not about yours. Discover the needs of your customers first, then start to think about your app idea.

You may want to accelerate the process, but that may not be a good idea. You should be aware of the challenges you may face, and that's the exact reason why I wanted to write this book.

There will always be a million other yoga apps out there. So, if you want yours to be exceptional, you need to think like your customers. You may not be able to give them what every other app couldn't, but that's not the point. The point is, you need to understand their situation and plan the details accordingly.

Then you'll need to validate your idea with your potential customers. Which may be easy if you're an extrovert, but it may be challenging if you're the opposite.

You should be able to live from challenge to challenge instead of paycheck to paycheck if you want to be an entrepreneur. So, be it. The more challenges you overcome, the more you improve yourself in your field.

How to Validate an Idea Without Development

In order to build a business, you need to find people who are willing to pay for the product you make. If you can't, it becomes a hobby, not a business.

* * *

To start with, you need to identify a problem. Then you can ask people if your product might solve their problem. If somehow it does, and can also deliver a monetary value, you may consider your idea validated.

You can start working on validation by asking random questions of random people. Remember, you need to validate the problem first, and then you can validate your idea. If the problem you think you have identified turns out not to exist, then your product is dead in the water.

Let's think about some different scenarios:

In our first scenario, say you want to build an app that contains nutrition information, but you don't know what problem it solves. In other words, you have an idea for a product, but you don't see the problem yet. Don't worry, we have all been there, it's the place where most people get stuck.

You can start by asking people about the problems they face in the nutrition field. They may want to scan the barcodes of a product and get some nutrition information. Or they may wish to enter details of what they have in the fridge so that an app can show them healthy meal recipes. The list can go on and on. That's why you need to find the problem first.

In our second scenario, let's take a step further. Perhaps you saw a problem in the meal delivery field. The problem is that sometimes students can't access healthy food, and some families need a little extra household income. You want to do something about that.

Then you realize, families may have leftover food after every meal. All you have to do is get the food from the families and deliver it to nearby students. That will solve the problem, right? It might.

Now you need to find a way to ask students and families if they really have that problem. Here comes the survivorship bias: You may ask a couple of people, and it turns out they are in that exact situation, but it doesn't mean you have validated your idea. It means you need to ask more and more people to make sure you have at least a decent number of people who have the same problem.

After you have formed a group, you can ask them about your solution. You don't need to share each and every detail with everyone, just give them a broad idea and enough detail for them to understand the concept.

Remember, you're seeking assistance here, you're not trying to convince people that they have a problem and need your solution. That's sales, not idea validation. Idea validation is making sure that your idea has an addressable market to begin with.

In our second scenario, you will be ready to analyze the data you have collected after you get the necessary answers from, say, ten different people.

The third scenario takes us a step further. You have seen a problem in yoga apps, and, as a yoga teacher with more than a hundred students, you have provided the validation yourself. You are sure you need an app to arrive at the solution, but you still need to validate it somehow.

In situations like lockdowns, you can't reach people in-person easily. So, you need an alternative, and it may be a better one.

Start by creating a form from Google's form building tool, or Typeform, etc. Enter some real questions about the problem and the solution you see, and try to find people to validate your idea.

If there are people who want to pay something for your solution, preferably upfront, then you can say you have validated your idea.

In your form, you asked people if they would pay for the app and, if so, how much. My advice is don't ask them about any "monthly payments" or "one-time purchases" if you don't have an exact pricing structure. Just ask them how and what they would be prepared to pay for.

Because they'll also give you pricing ideas if you allow them. Don't restrict them with a pricing strategy but allow them to think about your solution and how it fits into their lives.

Also, treat them like they may be your test users in the future. You not only need paying users, but you also need testers. The testers may not give you money, but they will save you some.

Idea validation is not an easy task, and you will need to work on it quite a bit. It's beneficial, though and it may save you from spending a lot of money on development and other such things.

Don't start development without validating your idea. All you need is an internet connection, and you're good to go. Don't hesitate to ask people for help. Remember, you might be solving one of their problems, and that's a good thing.

Also, continue to persevere with your ideas. You may not find the perfect problem/solution fit immediately, but you'll get there. If you listen to your future customers, they'll guide you through the problems and their solutions.

Eventually, you'll find some ideas to build upon, but take care not to fall in love with them at first sight. Ask people about them, communicate with them as much as you can. Then, you'll realize when you have a genuine, validated product idea.

Idea validation may seem harder and harder if you stick to "wrong" ideas. Remember, if you really want to have the app that no one else does, there may be a good reason why that may not be the best of ideas.

Think about the problems you can solve. If you look around, you'll see millions of people facing the exact same problems you have to address in your daily life. You can find a solution not only for them but also with them.

After validating your product idea, let's take a look at whether or not you need a designer and a developer. You may be more capable than you think.

* * *

How to Determine the Need for a Developer

First of all, working with professionals is almost always a good thing, so, I'm not going to advise you to do the opposite. Any work is done better with professionals, but the thing to determine here is whether your unique case needs professionals from the first day or not.

The prototyping phase doesn't need developers or designers, because it takes what you know and communicates it to those you show it too. Prototypes make everything clearer.

To make a prototype, you should find a way to deliver your product idea without coding. I wouldn't tell you to build a search engine without coding, of course, but you should find a way to deliver value with your idea without having to build it at this stage, which is about creating and sustaining your vision. Your vision may need a little (or more than a little) coding, but it's irrelevant when you're just getting started.

Start with the most affordable creative tools – a pen and paper. Simply write down what you want to build, just to clarify everything in black-and-white. In the upcoming chapters, we'll look at idea analysis, but that's not the point for now.

Once you have put your ideas on paper, draw your idea on another piece. I know it sounds tough, I can't even draw a straight a line, but you should find a way to do it. Try beginning with small rectangles.

Then connect the small rectangles with lines. Think of small rectangles as screens and lines as actions. Then fill the rectangles with smaller rectangles, thinking about how you'd structure your screens.

The above technique is called thumbnailing. It's used in prototyping as a tool to communicate the design to ourselves, not with the outside world. The fidelity level is so low that no one can understand anything from the little rectangles as they appear right now.

So, we need to increase the fidelity. This means you should draw the same rectangles, but a little bit bigger. This method forces you to think about what goes into which screen while not trying to make every rectangle pixel-perfect.

With that second step, you have just created your low fidelity wireframes. The fidelity is still low, but you can use them to communicate your idea. You can see what's coming next, right?

The next step is creating the medium-to-high fidelity wireframes, and they are the most critical wireframes you'll draw. If you're going to procrastinate, don't start here. Start at thumbnailing and stop it when you get to this point.

High fidelity thumbnailing is something which can change everything in a development lifecycle. You put what you want on paper and try to visualize every feature you have. At this point, you can get some help from brainstorming, if you hadn't already.

Using pen and paper is almost always enough to put something out there, but you'll eventually need to put those drawings into the digital world. You can scan them or take a photo of them to make them digital, or you can draw them on your computer.

There is a plethora of wireframing and prototyping tools out there for every platform, so you wouldn't even need a computer, as phones these days are capable of so many things.

After you have generated digital high fidelity wireframes, you may start to look for ways to showcase your product's features without coding. You may not find it easy, but you will find a way eventually.

You can use a pen and paper too, of course, but this time it's a bit different. For example, if you're building a product for a company directory listing, simply start by listing the companies. If you're building a photo filter app, try to replicate some of your filters within a design app on your computer or phone.

When you try to create value without coding, you'll soon realize why you need coding to build something. Technology is evolving day by day, and you should do your research well before starting something.

* * *

After all the research and the struggle to create your new product, let's say you found your reasons as to why you need development. It's now time to focus on the parts. The quality of your description of your project is directly dependent on your knowledge of its parts.

Believe me, there are (and always will be) more parts there than you can find, so keep on looking for details. The whole process described above is designed to enable you to find the details in your app that you weren't aware of previously.

You need to identify these details because someone eventually will, and if it's not you, it may be more challenging. Designing an app and writing its code is hard work, and you need to be confident the exact details are in your possession before you start work.

It's clear, the more details you have, the better. Let's build a scenario:

Say you want to build a photo sorter iPhone app which has just one task: sorting the photos you select by their dominant color. It may make your social media accounts more fluent.

How many pieces do you have right now? One, and that's the photo sorter app, right? Wrong. At first glance, I can see a whole heap of them. Like the following:

You'll need to build a dominant color algorithm. Finding the dominant color in a photo may seem easy, but don't forget you aren't looking for an exact color, you need to look for a "hue." The hue is the color's "temperature" or "tone." You may need to find a way to determine the dominant shade of the image.

Then you'll need a sorter. You may find millions of different color values after completing the previous step. Not every value you'll get will have one dimension, so you'll need to find a way to sort colors.

You'll need to implement those algorithms on the iPhone, which means you'll need to port them to the languages that can work on iOS (the operating system of the iPhone).

Next, you'll need to have a screen flow. It's like squares with names and lines between them (thumbnailing, right?). Because no one can develop your app for you without knowing what you want to see you'll eventually need high fidelity wireframes too.

You'll need to have a design of some sort, as without the design you won't get noticed by anyone. The design is essential.

Then, you'll need to determine whether you should have a server or not. It's about your thoughts about people's photos. You may want to have them for future AI use cases. (If that is the case, don't forget to ask for permission. No one wants you to use their photos unless they do.)

As you see, there are (and will be) more and more cases when you start to think about building your product idea. Even if it's a small one like a photo sorter, you'll have some details.

You don't need to be a coder or a designer to think like one. You just need to find a way to make every step of your product clear. In the modern world, you mostly need to assume the role of the product manager.

However, if you don't know what you're expecting, hiring a product manager won't make you happy either— but thinking about the details will.

The details of the product can also help you to find the developers you need because you'll be armed with knowledge and will know what to ask for. If you need extra help, though, you can always hire consultants (like me) to guide you through the whole process.

I've been doing it for years, so it seems easier for me, I know. But that's why I decided to write this book and tell all about the real process.

As I touched upon earlier, your conditions are unique and you need to find a way to use your conditions to your advantage. By all means, ping me if you need help.

Having determined the need for a developer, let's take a look at how you build your application flows and wireframes with real-life examples.

Creating Application Flows & Wireframes

We'll look at an example scenario here, and go through it step-by-step. We are going to build the photo sorter app, which we talked about earlier.

Taking our pen and paper, we start by drawing 1x2 rectangles. You can use any unit or proportions you like, they may be 1 in/cm wide and 2 in/cm high, or so. Just make sure that the height is bigger since we're talking about a mobile app here, and we're making it in portrait orientation.

We will have three small rectangles side by side. The first rectangle represents our home screen, where we can take different actions like selecting photos, choosing sorting algorithms, and placing in order.

The second screen is the photo selection screen. We'll tap a button on the home screen and come to this one. So, we need a line in-between them.

The third screen is the sorting screen, where we'll see the photos sorted; we'll be able to change the sorting algorithm or even change the order manually. We'll select photos from the second screen to fill this one. So we need a line here too.

Now, let's make them slightly bigger. I'd say 3 to 5 times bigger should be enough. We keep the lines between them and start filling them with components.

For the first screen, we'll have a grid of buttons representing different sorting algorithms and types. We'll also have a button for selecting photos. You may want to utilize a help screen somewhere, but to keep it simple I'm not doing that here.

So, now we have a four-rectangle button grid above and a rectangular button below. Let's draw them on the first screen.

* * *

Users will tap one of the sorting type buttons, then tap the button below. That'll bring them to the second screen Which will be filled with a grid of rectangles. There may be some buttons for approving the selection, going back, and resetting to make a new selection.

Users' photos will appear here. So, we'll need to have permission from them to use their photos. Keep that in mind.

Users will select the photos they want and tap on them, therefore we should make it clear that the images are tappable. After they tap on them, the images will be selected. Users may select as many images as they want.

Then they tap the "Done" button, so we need to draw one. After tapping it, they arrive at the sorting screen. There may be a loading spinner and it may take some time to sort that many images, but we don't need to draw that right now.

In the sorting screen, images will appear inside a grid. They appear sorted, but because we want users to change the sorting type, we have a grid of photos and a type changer button. Again, for simplicity's sake, at this stage we won't get into the type changer selector screen or other such things.

Now we'll give users the ability to change the order manually by dragging the photos around freely. Then, as we need some way to save the sorted photos, we'll use the device's folders and proper naming conventions to do so.

Here we have a complete application flow. Well, perhaps not complete, but simplicity is our watchword here.

In the next step, we'll draw wireframes.

Wireframes

Wireframes can be digital, so you can use various apps to create them. Wireframes are sometimes also called app mockups; an extra bit of information to help you in your search for a tool.

* * *

Since we want to create an iPhone app, our wireframes will be the size of an iPhone. I know iPhone sizes vary, but just select one from the middle range. We'll have three screens and we already know what to draw on them.

We'll actually draw more than three screens, we'll also draw the states of the components there. Like a button. When you think of a button, you think of at least two states, pressed and non-pressed. If it is hard for you to illustrate what a simple button's pressed state is with just a pen and paper, then it's not essential.

You just need to know there are different states of the components and if you want to communicate them you will need to draw them somewhere. We'll also utilize words in our wireframes because of two things: 1- We may need text on the screens 2- We may need to annotate something.

Annotations are important because they can save you from drawing anything else. So, if you don't want to illustrate a button's states, just write something about the states nearby.

By using this method, you can start to draw your high fidelity wireframes. They'll be an integral part of the communication between you and the outside world.

We'll get into real designs soon, but meanwhile, wireframes are important in communicating your ideas to designers. Even if you're a designer yourself, wireframing makes it easier for you to iterate your designs before you spend a lot of time crafting the details.

Here's a quick recap before we dive into the design & development world:

We started with thumbnailing. When you have an idea, you can just start there; no analysis necessary. Thumbnailing may illustrate the big picture in which your app lives. You'll find out what features you need to implement from a bird's-eye view.

Then we continued with slightly bigger thumbnails. They gave us the power of both speed and detailed thinking. It's like preparation for the wireframes. Because when you start directly with wireframes, you'll have to draw a lot more than is actually necessary.

Next, we came to wireframes. Wireframes are the main communication tools we have with the designers, developers, and even users. Remember we talked about validating your ideas earlier. You'll need something to show to your wannabe users to get feedback from them.

Of course, in real life, there are almost no three-screen apps, but the point is that we need a system to improve. You can even make a fifteen-screen photo sorter app and it may prove to be a better execution of this idea, and it's totally fine to do so.

As I mentioned earlier, your conditions are unique and you'll find your own way of doing these things. I'm just giving you some initial guidelines.

Let's have a brief look at how to work with designers.

How to Get Your App Designed

Design, as a term, can have multiple definitions. It can be summed up in one sentence or it could be as long as an entire book. From our perspective, the design is how things look, feel, and work.

How do your product, branding, and marketing materials look, how do users interact with them, and how do they feel about those interactions? All of those questions are about design.

You can either learn how to design something by yourself or find a professional to do it for you. Remember, it's a profession, and doing it right is hard. You may have the ability to appreciate good design, but that doesn't necessarily mean you can create it.

Let's take a look at app design and product design in general, and then we'll get on to finding designers and working with them.

* * *

What Is App Design?

App design is a small part of the broader term "design" that we'll use. It's about how your app looks and feels.

The app design has several parts that you need to think about.

First, the layout. That's why we created wireframes. Everything gets smaller when it comes to mobile devices, and you need to find a way to put over the information you want with limited space.

Second, interactivity. There are too many boring apps out there. If you don't want to be one of them, you need to find a way to stand out. And it's not just about the colors and wording of your app. It's also about interactivity.

Interactivity is mostly the "feel" part of the design. How your buttons react to users' taps, how your app changes screens, and whether it uses haptic feedback or not. These all are about interactivity.

Third, colors. Screens are getting better and better every day, but they still have limited color space. This means you can't use every color you can see in an app, you need to select the colors wisely.

Also, you shouldn't use too many colors together because it confuses people. You should find a color palette and stick to it.

Let's talk about what you should have on board to allow you to say you have a complete app design:

What Are the Deliverables for an App?

An app design is not just the app itself. There are several elements that you can't make your app live without, and here are the things you need:

1- App Icon

There is no mobile app without an app icon. Your app icon is the thing users will see first. Technically, they're squares, and they need to be bigger than 1536x1536 pixels.

You'll need your app icon to be resized for every possible use case, and it's so much easier if you design big and then scale down.

2- App Screens

Screens are the parts of the app where users will interact with. Screens may be designed in a 1x scale, but they'll be scaled to 4x and beyond. Therefore, you should design them in vector form, not rasterized.

The main difference between vector and rasterized design is that rasterized graphics are made of pixels, while vector graphics are made of paths. This means vector graphics can be sized to virtually any size possible.

3- App Interactions

This may be called an app prototype. Modern design tools offer you great workspaces for converting your designs to prototypes. It's not critical, but it's a nice thing to have. If you can plan your interactions before the development phase, your developers will thank you. We'll see why in the upcoming chapters.

4- Store Screenshots

App Store, Google Play Store, or any other store out there will require your apps to have screenshots. You can also include videos. They'll be the size of a screen and you can have more than a handful of them on show.

I use the term "screenshot," but they aren't necessarily real screenshots. You can design them however you want, but make sure they represent your app.

The store screenshots need to be as carefully crafted as your app design because they'll be the things which persuade users to download your app. You can even have different versions of your store screenshots which you may test from time to time.

5- Icons

Not the app icon, the icons inside your app. Almost all apps have icons and some raster graphics within them. You should convert your icons to raster graphics and make sure they're available in all necessary sizes.

* * *

Why not just use the vector version? Because rendering small yet complicated vectors is harder than rendering a bunch of pixels for a phone.

Now we have arrived at the part where you start interacting with designers in the real world.

What Will Designers Expect From You?

No one will give you what you want if you don't tell them what it is. You need to find a way to make sure they understand your vision when it comes to your product.

That's why we created wireframes. They're the main tool we use to communicate our ideas to designers. But that's not all.

Designers will need your product's brief, indeed, everyone you work with will need that the brief. A brief is basically everything you can write about your app in a detailed overview. I'm emphasizing this here because there shouldn't be any technical details.

You'll try to communicate your vision using words. You'll also need to describe the feeling you want to give to your users.

What to Expect From Designers?

You'd expect them to deliver the app's deliverables, of course. However, if your input is useless, the output will be useless too. Ultimately, if you don't have a great product brief, you can't expect a great design.

In creative tasks, being restricted brings about creativity, so you should set your restrictions as clearly as possible. If you need advice about this process, you can always hire consultants (like me) to guide you through the unique flow of your project.

You may end up with a great design without providing much information up front, but most of the time it won't be the thing you want. Having a great design is completely useless without a use case; think thoroughly about your use cases.

How to Find Designers?

You can find designers on freelancer or portfolio websites, local communities, etc. However, you'll notice the price range is broad.

You need a way to grade designer portfolios, and this takes time.

A better approach may be collecting the designs you like on a mood board and using it to describe your product to designers and get quotes.

If you want to work with high-quality designers, but your budget is limited, you can always work with consultants, such as I, who can connect you with some great designers. This will come with a bit of an upfront cost but, based on my experience with clients over the years, it makes the whole process significantly easier.

Say your app is now designed. Then you'll need to find some developers to develop it, right? But what kind of developers do you need? Those who are building robots? Let's take a look.

What Kind of Developers do I Need?

Yes, there are different kinds of developers. By saying different kinds, I mean different professions. Development is a profession by itself, but it has many parts, and no single developer knows everything about every one of them.

If you need different parts developed for your product you may also need different developers to develop those parts. In the same way that your product has different pieces there may be different developers who are experts in each of those pieces.

The thing to do here is determine what your needs are and how you can make the pieces work together.

* * *

Let's take a look at different types of development work and discuss the details:

Mobile Development & Mobile Developers

Mobile development is development for mobile phones and tablets. It's a broader term and it has three main parts: Native iOS Development, Native Android Development, and Cross-Platform Development.

IOS development consists of iPhone and iPad development. Lately, Apple has made macOS development almost the same so that you can add a bit of macOS development into it. There are two programming languages widely used to develop iOS Apps: "Objective-C" and "Swift." Both are made and maintained by Apple.

You can also write code that works in iOS with lower-level languages, like "C" or "C++," but it's not that common.

We'll take a look at each language in future chapters.

It makes sense that if you need iOS Development, you need to work with someone who can write Swift, or at least Objective-C.

Next, here comes Android.

Android is a platform from Google, but they didn't create the necessary languages: "Java" and "Kotlin" are used intensively to develop Android apps. Android apps are also used in any kind of device which can run the Android operating system, so your device range is dramatically broader.

Just like iOS, you can develop Android apps with lower languages like "C" and "C++" too, but, you guessed it, it's not that common.

If you want your Android app to be developed, you'll need developers who can write Kotlin or Java. They are interoperable, so both may work in the same way.

* * *

Up next, Cross-Platform mobile development.

It's a highly debated field. There a is a multitude of tools and languages which try to claim that they work as well as their native counterparts, but they have limitations.

As I said, there are so many of them, but here are the most popular two: "Flutter" and "React Native." These two are frameworks, not programming languages. We'll discuss frameworks in the later chapters.

Flutter is a relatively new one, developed by Google upon their "Dart" programming language.

React Native is developed by Facebook, upon the language "JavaScript." Note: they're entirely different languages from "Java."

We'll come to the native development/cross-platform development debate in the upcoming chapters, but, for now, just be aware that they exist, we'll cover the rest.

Frontend Development & Frontend Developers

Frontend development is the process of developing interactive user interfaces for web and mobile web. By web, I mean the websites you can open with a web browser. By mobile web, I mean the websites you can open with a web browser on a mobile device.

Frontend development, as a concept, is relatively new. All frontend development processes consist of three languages: "HTML," "CSS," and "JavaScript." Yes, the one you can also make cross-platform mobile apps with, not the one in Android development.

HTML stands for "Hyper-Text Markup Language," and it's used to get things to a web page. It consists of components, which we call "tags," and it communicates what will be on the screen.

CSS stands for Cascading Style Sheets. It's used to style the HTML elements. You can do pretty much whatever you want with it, so the creative part is firmly on CSS' shoulders. It's not a programming language by definition, rather it's a language to give some direction to styling.

JavaScript is used to make the previous ones interactive, at least that's its use in web development. In modern times, as data processing and transfer became cheaper, we introduced frontend development frameworks.

There are a ton of JavaScript frameworks, but the most popular are React, Vue, and Angular. We'll get back into that.

Backend Development & Backend Developers

Backend development is about everything you don't see at first glance but it's all necessary for your product to exist.

It's the code running on the servers. A server is a computer located in a datacenter somewhere in the world. You can also put servers on your local computer. You can run code on those computers and interact with them in several ways.

You can develop application backends with almost all the programming languages out there and there are no limitations. Since there are no limitations, you'll need to limit yourself.

You'll need to limit yourself to your particular use case. The most common backend development languages are, "Java" as you saw in Android Development, "Python" (not inspired by the snake, but the show "Monty Python"), "Ruby" (we call packages "gems"), and "PHP" ("Personal Home Page").

Of course, you can use different languages for different purposes too. For example, you can use the "Go" language if you need high-level (easier) coding with low-level (faster) performance.

And, here's the newcomer: AI.

* * *

Artificial Intelligence Developers / Data Scientists

Data science and artificial intelligence are relatively new areas of development. The reason why they're relatively new is the decreasing cost of storing and processing data. The current level of technology allows us to analyze and predict large quantities of data, and this ability gives us a vision of what the future will look like.

Data science is searching for some meaning within seemingly meaningless data. Hence, the field of data science is a field of research-first, not development-first. However, if you want to develop a bitcoin predictor app, you'll need some data scientists along the way.

The languages "Python" and "R" are the most commonly used languages in the field of data science, but you can do the same tasks with almost every language out there if you have the time.

Let's talk about quantity.

How Many Developers do I Need?

Short answer, at least one. Long answer, it depends. Let me explain.

You'll need someone with development skills, but they may not be developers. Like yourself, you can learn coding and start developing your own projects. It may work if it's a side project, or you have enough time to do it all by yourself while learning.

However, when it comes to releasing a new product, there will be countless topics that need to be covered, so you'll be pretty busy with all of them. At least you should be. Otherwise, you may have a hard time acquiring users.

Of course, there are exceptions. I was one of them when I started coding 15 or so years ago. I wanted a website to be developed and as I didn't have any extra resources, I developed it myself. Then I realized that was a great journey of learning and exploration.

But, there are always exceptions. You don't need to be one of them.

There may be survivorship bias here, so think carefully about the real needs of your product in its development.

Here are three scenarios for you to think about a bit more deeply. In no particular order:

Scenario 1: Photo Sorter iPhone App

Yes, the beloved photo sorter app. We already have an idea about what it is, so we'll need an iPhone developer to develop the iPhone app. It will be better if we can find one with the necessary skills to develop core photo sorting algorithms.

If we can't, there is another way. You can go to a freelancer website, describe your project in broad terms, and wait for the quotes. After you get the quotes, you might find someone who can code the algorithms with the Swift programming language.

You'll have two options: First, you can give the entire app project to the one you found on the freelancer website, but they may be more expensive because they have the skills to deliver the whole thing. Second, you can get the algorithms developed by them and give them to your lower level iOS developer. This may be cheaper, but having different people working on the same project is not always a good thing. I'll explain this concept later.

Also, you may not need an iPhone developer if you want your app to be developed as a cross-platform app. You can find a Flutter or React Native developer, and you'll have two apps, one for iOS and one for Android.

However, there is a caveat. Cross-platform development is still not as performant as it could be. That's because you'll have another layer on the devices' operating system, and they are not designed for that. They are designed to be utilized in the way they want, not the way you want. We'll also cover this topic in more depth later on.

You may also need a backend developer if you want to store the images and user accounts. I said you may, because there are some platforms like "Firebase," which can act as your application's backend without writing code (or writing just a little).

* * *

You'll need to decide whether if you store the data or user accounts. The project we described in the earlier chapters may not need those things. However, there's always room for improvement for us entrepreneurs, right?

Scenario 2: Early Days of Instagram

Another photo app, and an extremely successful one. Instagram was launched in 2010, when there were no cross-platform frameworks in which to develop a photo filtering and sharing application. Instead, they wrote a native iPhone app with a Python (Django, as the framework) backend.

Instagram's founder was an exception, too; he learned to code at night to develop side projects. Then his side project became Instagram. We'll take a look at the development angle:

Say you want to write an iPhone app which you'll use to filter and share photos. This means you'll need at least one iOS developer. In those days, we were writing iPhone apps with Objective-C, but now you can use Swift as well.

Photo filtering is an easy task these days because there are numerous different frameworks which have been developed around it and they do the heavy lifting. But that was not the case back then when more development power was needed.

You probably won't need that much help. Nowadays, one iPhone developer can develop a photo filtering and sharing app.

If you're not willing to learn backend development, you'll also need a backend developer. The backend developer will be responsible for keeping user accounts in sync, saving photos to the cloud, and managing the social interaction in the backend.

In conclusion, you'd need at least two developers, and less is not more here.

Scenario 3: Stock Market Analysis & Aggregator Website

In our third scenario, say you want to aggregate stock market data, analyze it, and make predictions about it. Your channel will be a website and you want interactive charts and suchlike to display data and predictions in real-time.

Starting with the basics, you'll need an AI model to predict stock prices. It may be developed using "machine learning" techniques, but using "deep learning" techniques may be a better fit for this application.

You'll need at least one data scientist who can develop deep learning models. The more data scientists you have, almost always, the better.

Then you'll need a backend developer who'll get your purchased stock data b onto your servers, make the predictions using the AI model, and give an "Application Programming Interface (API)" to the frontend developer to use.

Next, you'll need a frontend developer who'll show the data you have to your users. They also build the look and feel of the application via code. Building the graphs, working in real-time, drawing shapes in the graphs, etc. Such is the job of the frontend developer.

You may need at least three developers to build this website in normal conditions.

How Many Developers Do I Need?
At least two developers working on the same project makes it more sustainable because writing code for someone else is better than writing code for yourself.

However, you may not have a budget for that. In conditions like these, you'll need to think about which parts are needed urgently to generate capital. After you have spotted these, you'll need to find someone who can do those particular things well.

The more clearly you define your product, the easier you will find the search for developers.

* * *

How to Reach Out to Developers

Finding developers is not an easy task. You might have developers around you, or you may know a number of them but that doesn't necessarily mean you will find a perfect fit for your project.

Here are some places where you can find developers:

1- Friends & Family

There may be developers in your friend or family group. That doesn't necessarily mean that they'd want to work with you, but you'll always have someone to ask something development-related within your close circle.

Friends and family may not be a good fit, however, because they will be somewhat biased about your work. Whether positive or negative, being biased is not a good thing when you try to convert friendship into a business partnership.

2- Freelancer Websites

It may be easier for you to find someone by reaching out beyond your immediate network. You can always go to a freelancer website and post a project listing; then, the quotes will come.

However, there is a caveat. This is a way to find freelancers, but you probably won't have the necessary tools to assess them. This means a freelancer may have been a good fit for a thousand previous projects, but it doesn't necessarily mean that they'd be suitable for your project.

It would be ideal if you had some way of gauging who best to work with, but that comes with experience and might cost you some money. Don't worry; sometimes learning takes time and money.

3- Developer Facebook Groups

Because there are Facebook groups for almost anything, there may be a group which fits your particular situation. Let's say you want someone to build you a website with "Django," a framework written in Python. Simply search for "Django Developers" groups and join them. After you have joined, share a post with a brief introduction to you and your development need.

The proviso here is that with freelancer websites you at least had user comments to guide you, at least to some extent. With this method, you're on your own. You don't have a way to determine whether they're an excellent fit for your project or not. So, you'll need to interview with them personally or ask for help from someone who is experienced in the field.

4- Developer Email Groups

Believe it or not, there are still email groups out there, and there are developers who are active users. You can find out more about them online and start by sending emails to the shared addresses.

It's a bit of a shot in the dark and you may not want to follow this route at all. But you never know, the knowledge that they exist might come in handy.

5- Developers From Abroad

You can always search for foreign developers in other countries. It's especially valuable if your country's currency is stronger than the other country. This basically means you can work with someone without paying them as much as you would someone in your own country.

However, here comes another caveat: They may not be governed by the same rules that apply in your country. So, in a conflict situation, you may not have a leg to stand on.

There are ways to avoid this kind of situation, such as using a platform for making and receiving payments, and handling disputes.

* * *

6- Getting Help

You can always get help from someone who is familiar with the development business. Someone who can grade developers, interview with them, source developers from abroad, and form you a team if necessary.

Some people and companies do this kind of thing as a job, and I am one of them. I help entrepreneurs like you to find developers and guide them on the development journey. Whether you're just starting or you're halfway there, let me know if you need some help along the way.

Tips & Tricks on Finding Developers

Here are some tips and tricks on finding developers and working with them:

1- Have Someone to Trust Right From the Start

It's the first person you hire. It may be a consultant, a CTO, or a business partner who understands these matters better than you do.

This person will be responsible for the formation of the development team and helping you understand each other when talking about technical topics. Because your terminology may be different, it's a job that's akin to translation.

2- Have a Solid Brief at Your Fingertips

Know what to ask, and know it well. You'd want to have the answers when they ask you questions, so you should be able to pitch your project to developers to work with them successfully.

Because they'll also try to find out whether they can or cannot code your project and if it'd be a good fit for them or not. It's a two-sided process, and one that can sometimes be hard for both parties.

* * *

A solid brief solves many things all at once. For example, it solves the question/answer pressure. If you have a brief that answers the tough questions, you don't need to answer them time after time.

3- Start With Small Tasks

When you find someone to work with, don't give them everything right from the start. You may want to start gradually, because whoever you are working with, you don't know at first if you will fit together well.

You'll need to find your way together, step by step. Start with small development tasks and make sure they're working as advertised. If not, you'll be able to disengage and stop working with them without losing much money.

4- Don't Stop Searching Even if You Find Someone

Finding someone to develop your project doesn't necessarily mean you won't ever need someone else. Things change. Your project will evolve. You'll need someone else someday, for a similar task or a different one, it doesn't matter.

If you don't stop searching for developers, you'll eventually have a pool of contacts and, when the time comes, having some names in the bank makes everything easier.

Take note of who does what and who can demonstrate their skills. So, don't stop searching and don't stop interviewing developers. Of course, ethics dictate that you should explain your conditions before you interview them.

5- Don't Stop Learning

As a tech entrepreneur, you don't have the luxury of not having to learn new things. The Tech world is fast and you need to keep pace with it. It's not necessary for you to learn every technical detail out there, but you should have some idea about all the things that matter to your business.

The more you know, the more you understand what's going on around you. Especially in development, if you know the necessary parts, everything else will fit together. Don't forget to read and share what you have learned with the global community.

The world needs developers and it also needs problem-solving entrepreneurs. The more knowledge both parties have, the better the world can become.

Chapter Recap

Let's recap the chapter before we move on; to emphasize the parts that I found important:

1- What to Do Before Planning Your Next App

In the first part, we talked about your unique conditions. It's important to know who you are as an entrepreneur; that's why I suggested you check your immediate status first.

After establishing your exact situation, you can navigate a path towards executing your ideas. Also, determining what you can afford by yourself and what kind of funding you may need to find is an excellent way of understanding the financial situation.

After you have determined your financial position, ask people about your idea, and try to determine if there's a real need for the product you want to build. That's important, because if there's no need, you have no chance of success.

2- How to Validate Before Development

In the second part, we talked about validating your idea. You may want to build a product, but you still need to sell it. If you can't sell it, it becomes a hobby, not a business.

* * *

Validating your idea is about asking questions while having some solutions in mind. Trying to validate your solutions without binding yourself to them. You may change the solutions slightly or drop them along the way; all of those moves are improvements. You just need to know if your solution might work.

Knowing your solution works is simply about finding someone to pay for it, or at least someone who is eager to test it.

3- How to Determine the Need for a Developer

In the third part, we talked about creating your product's prototype without coding or doing any designs. It starts with a pen and paper (you can go digital too, of course).

You can start by drawing what's in your mind. You can use prototyping and wireframing tools; there are plenty of them out there. Drawing small boxes will help you see what the application flow looks like.

Then, it would be best if you created higher fidelity wireframes with more precision and detail. When that's done, the details of your product idea become more and more precise. Don't think about perfecting anything from the start but don't be afraid to redraw the same screens repeatedly.

4- Creating Application Flows and Wireframes

After we talked about the theoretical parts, we started to play with an actual scenario in part four. We created the app flow of a photo selection iPhone app and then made wireframes for it.

Going higher fidelity makes it harder at first, but it'll be beneficial when you want to communicate your idea to someone else, especially with designers and developers. They are the ones who'd need to know what's on your mind to make it real.

Here comes the bit about communication:

* * *

5- *How to Get Your App Designed*

In the fifth part, we started by looking at the term 'design' and how you'd use it in your product development workflow. App design (or web design) is the service you might need, especially if you're a tech entrepreneur.

Design is about how your product will look, feel, and work inside and out. The app icon and screen designs should match the mood of the store screenshots and the marketing campaign images. It's the same whether you're building a mobile app or not, you'll need design somehow.

After we talked about deliverables, it was time to determine what to expect from designers and what to give them. You should give them enough information about the idea, and the more details, the better. It would be best if you described what using your product feels like.

6- *What Kind of Developers Do I Need?*

In part six, we took a look at what kind of different developers may be needed in different scenarios. There are more development tools and types of developers out there, of course, but we just want to make sure we cover the initial, basic scenarios.

In short, you can develop a mobile app in two different ways: Native and cross-platform. Native means you code with platform-specific languages and patterns. Cross-platform means you code once and use it everywhere.

It seems like a no-brainer, but there are caveats; you can't get native performance with cross-platform development.

In frontend development, there are numerous frameworks, but they all are written in three languages: HTML (to get the data to a web page), CSS (to style it), and JavaScript (to make it interactive). The only real programming language among these three is JavaScript.

There are also countless languages and frameworks in backend development, so you'll need to decide what you need. If you need any help here, please don't hesitate to ping me.

7- How Many Developers Do I Need?

In chapter seven, we tried to determine how many developers we might need in three different scenarios.

In short, every part of your product needs at least one developer. There may be developers who can code all the parts, I know because I'm one of them, but it'd be better if you could afford more than one.

8- How to Reach Out to Developers

Reaching out to developers is easy, but finding the right one is not. So, in part eight, we took a look at how you can find developers who meet your requirements.

In short, every social circle you can try works in some way. Your friends and family may help you find someone to work with, or you can utilize listings on freelancer websites. Developer groups on social media may work for you, or you can ping me, and we can look at finding a good fit for you.

9- Tips & Tricks on Finding Developers

In the last section, I gave you the five tips I use the most. The most important point: Do your homework thoroughly and you'll be good to go.

Remember tip five, don't stop learning. Learning is the ultimate pathway to growth and development.

In the next chapter, we'll take example projects and prepare them for development phases.

Let's begin.

CHAPTER TWO

Before the Project

Idea Analysis

After taking a look at who to work with, let's focus on shaping your idea and making it easier to develop. We'll have three analysis elements: Idea analysis, business analysis, and software analysis.

We start here with idea analysis.

What Is Idea Analysis?

It's not enough that the only one who knows about your idea is you. It may also mean that even you don't know the exact idea either. That's because ideas are living things. They grow with sharing and reshape themselves.

So, before share your idea, you'll need a way to develop it in your mind first. That's more like writing than thinking, by the way. Everyone needs to be on the same page for your product to work correctly; it eases the information flow.

There's a personal element in idea analysis and yours will contain your personal qualities as well. That's why every execution is different, even if the ideas are the same.

* * *

1- Spotting the Details

We start with finding out the details surrounding our idea. You may already have the wireframes and application flow, so you can start there.

Begin by finding your application's revenue streams. Because you'll need to make money, right?

Then, continue with the user's journey in the app, from start to finish. It's called user flow, and it's essential. From opening your app (or website) to closing it, you can include each and every detail about what your users can do with it.

Write every step down on paper. It'll help a lot in clarifying users' expectations for you. You need to know what users expect from you so you can deliver maximum value.

2- Creating Personas

Identifying who your users are is an important task. Your users are the ones who pay for your app, so it's best if you know as much as you can about them and their characteristics.

For this very thing, we use personas. Persona means a descriptive document about one person that represents a group of people. You can utilize as many personas as you like. However, here, less is more.

You should know who your users are. So, start by giving them a name. Let's call this one, "Sarah," for example. Sarah, a student, aged 24, hustling a little extra cash as a social media marketer, and maybe in need of a photo sorter.

The more details you include, the better. You'll try to find people like Sarah in the future to use your app, so include as many details as you can.

Jane may not be the perfect fit for your app; by the way, that's why creating personas is essential. You can test the personality traits of your users on paper without taking any risks.

* * *

Even if you have just one laser-focused customer segment, it'd be better to   include more personas. You can make personas from different professions, different life challenges, and so on.

3- Using Your Users' Language

Your users will be the ones who make you successful. So, you should be able to speak in their language. By your users' language, I mean including phrases they use in real life in your app itself and its marketing copy.

Using personas to find out who your users are makes it easier to reach and interview them if they're different to you. You can utilize those interviews to learn about your users' day-to-day language and think about your application flow from another perspective.

For example, the color red may mean different things in different countries. Some may see it as love; for others, it might evoke blood or something else. As I said earlier, getting to know who your users are is the most critical thing you'll get to do.

Let's take a look at the developers' language.

4- Creating User Stories, or, Using Developers' Language

Here we are, on the first touchpoint with real development. User stories are one of the most important things when it comes to the software development process. They describe your users' journey in a structured manner.

That's a vast improvement on the idea analysis and it will be a smoother process if you include sufficient details.

Let's see how the user stories are structured and what they mean for developers. Here's an example user story:

"As a user, I want to sign in to the app with my email and password so that I can persist my data using my user account."

* * *

User stories are structured in three parts.

The first part, "As a user," means that "I'm using the app as a regular user right now and this is my story."

The second part, "I want to sign in to the app with my email and password," means that the "who" will make a "what," and the application will have to react to it in some way.

The third part, "so that I can persist my data using my user account," is the benefits part. That's "why" the user is doing the "what", and that's why it's essential. It also describes how important this particular user story is.

Let me explain what a developer sees in the first two seconds when they look at this same user story. This is what this user story means:

1- The app will have users who are called "users". It seems obvious, but that's the level of detail in the user story. There may be different types of users in the future, so I have to make it clear.
2- We'll keep user accounts, so we'll need a database somewhere. We'll have at least two fields to fill in: email and password, and we'll need a way to store passwords securely.
3- There will be a thing called "user account." It may contain more fields in the future, so it needs to be extendable.
4- There will be a sign-in page to be designed and developed. That'll bring a sign-up page and a forgot-password page in the future.

So, as you see here, describing the details is essential. Someone will eventually have to describe things like the sign-up page or the forgot-password page. You should take responsibility for thinking about such things if you want your experience with developers to be a breeze.

Let's talk about the next step, business analysis.

Business Analysis

Business analysis is the practice where you can see and maybe solve business problems upfront. You can think about it together with idea analysis, but I prefer to keep them separate. They both have some common properties, but they consist of different disciplines.

Business analysis is more like having a business plan for the software development parts of the business.

1- Defining the Workload

Defining the workload is one of the critical parts of business analysis. You may be thinking that you already know your product's boundaries clearly but writing about them makes communication easier.

It would help if you start by writing down the bare bones of the product you want to build. First, write a mission statement and a vision statement. These are essential parts of your business because, if done correctly, everyone to whom you communicate your ideas will know what you're trying to do with your product.

Then, define what you need to have within those statements. In short-, mid-, and long-terms, it's better to have an idea about what your product is going to be like. That's because, in software development, developing for extendibility is essential.

Even if you have a clear set of directions, there'll be many ways in which your exact idea could be developed, so one of the things you'll need to decide is if you want to extend it in the future or not.

For example, if you just want to have a photo sorter app and you're not planning to extend its functionality, there will be a different workload than if the opposite was true. Extendibility is a software quality, and as it takes time, it helps when you know what you want from the developers.

2- Putting the Workload in Pieces

The next part is dividing the workload into pieces. It's a bit harder, especially if you're detail-oriented, because there will be numerous details involved in this step.

Let's look at the same example: our beloved photo sorter app. After you've decided you'll extend the functionality in the future, it's time for you to set out the details.

To show you the level of details, let me give you an example:

Photo sorting functionality in the app will have four different key points. It'll sort the photos via the dominant color, average color, dominant hue, and average hue. To find those key points, the app will first analyze the photos, so the users will have to wait a bit before the app does its job. This means we need a loading screen or modal panel to show what's going on.

By including this kind of detail, we can safely say we're on the right track to keeping our promises to users—the promises we made with our mission and vision statements.

The pieces you should have in hand are like these, and they're easier to produce when you're with someone. It may be more challenging when you're alone because it works better as a brainstorming session. Putting ideas into their places may be easy at first, but making them smaller and smaller gets more challenging with time.

3- Creating a Market Vision

You may be asking, "What does this have to do with software development?" Let me explain: Other companies and products may do what your product will do, and taking a look at them doesn't hurt; it helps.

There may be a feature in one of the competition's apps that your users would love. Or there may be a problem with one of these apps that you know you can solve with yours.

You should know your market before building your idea by downloading all of your competitors' apps, using them, and noting every feature they have before you even start.

* * *

Then, you can start grading those features. Grading is easier when crowdsourced, and you may already have a crowdsourcing channel: comments. Comments written about the competition may enlighten you about your users.

Think about that. Before Instagram, almost no one knew they might need something like photo filtering. Now there are a million different photo filtering apps which all claim to have unique abilities. Fun fact, they may even have such abilities.

And here comes the critical part about reading comments. Don't believe them; just analyze them with insight and empathy. That's all you need.

4- *Thinking About Future Problems*

Thinking about the future is hard. However, this part is one of my favorites. You'll have problems no matter who you are and what business you're in, so thinking about them won't hurt you; but not thinking would.

One of the future problems may be in the logical use cases of your app. It may be the thing the world needs right now, but it may not be the same in the future. I'm not suggesting you back up; I'm just highlighting it as something for you to think about should it occur and consider what your reaction might be.

Because problems will happen. For example, you may not be a perfect fit with your developers, and if you use equity sharing but don't utilize vesting from day one, you'll lose shares in your company as a result.

You might spend an extended period thinking about that kind of problem, but I suggest limiting this timeframe. Because thinking about problems can lead to frustration, and that's not what we want. Limit yourself to two hours, for example. In those two hours, write down the potential problem areas. You'll solve them eventually.

* * *

If you need help, you can always ping me or other professionals who can help guide you through the entire journey.

5- *Preparing For Software Analysis*

Preparing your idea and business analysis for the third part is the easiest one. As a tech entrepreneur, someday you'll need to place your trust in developers. Trusting developers is a good thing, but it doesn't necessarily mean you don't have to do anything.

Preparing for software analysis means writing it all down somewhere. It may be a physical or digital piece of paper, but you need to apply yourself and do the writing.

Everything we've discussed so far is better when written. You may have all the ideas and analysis in your head, but it's hard for you to explain it all again and again. Written communication makes everything clear when explaining a bunch of ideas to yourself, and the same goes for the rest of the world.

Software Analysis

After completing the idea and business analysis, here comes the third part of making your lives a bit easier when it comes to speaking the developers' language. We're talking about the software analysis document.

Traditionally, the software analysis document is a highly specialized one, packed full of technical details. However, in modern times where tech startups are moving quicker than ever before, it has become less technical.

I suggest you find someone to consult with before and during the creation of the software analysis document. Because we're going to try to make developers stop thinking.

There's an interesting concept here. Software development is all about thinking, right? Well, sometimes. But thinking will be done before coding. If someone tries to think while coding, it slows them down.

In the same way that Shaolin monks go about things, coding is a muscle which needs training, and the part involving typing is the one to do without thinking. Any decision about the code to be written has to be made beforehand.

Otherwise, the whole process becomes slower and slower with every decision made throughout the development phase. That's mainly because your decisions affect other parts of the software, and the other parts become more prominent and more complicated with time.

So, we need to make timely decisions beforehand to ensure a perfect document is ready to communicate to the developers. However, you guessed it, there is no such thing as the perfect document. So, we'll try our best and get help along the way if needed.

The software analysis document may have some structure, but what you can put in it is not restricted. So, as the first rule, we'll put as many details as we can into the document ensuring every known detail is included.

Let's think about our photo sorter app scenario:

1- Business Processes

In the first step, we start by listing our business processes. Business process means the things your app and users can do together, like having a user account. When we write something about having an account in the software analysis document, it should include every detail we know about the authentication and authorization process.

For example, the methods we can use to authenticate the users are essential. If we use social login, we should include which platforms are to be used as a login provider. If we authenticate the users via email, we need to decide if we want to validate their email by sending a confirmation message.

* * *

These may seem like no-brainer questions, but we're talking about people talking to machines here, and if you want to talk to machines successfully, having crystal clear instructions to follow is key. Everything we include in this document is designed to make the developers stop thinking about additional things, and it eventually speeds up the process.

From a business perspective, validating users' emails may be a critical thing, and everyone should know that, right? That's irrelevant when it comes to the development of that exact process. Developers develop what you decide is to be developed; it's not their job to think about the business process as a whole.

If we say, "users will have user accounts" and nothing more, it means nothing to a developer in the development phase. However, it brings a number of questions to their mind, and we should try to answer those in the software analysis document.

2- User & Application Flows

Previously, we talked about application flows. However, in the software analysis document, the application flow is a bit different. We'll try to make developers understand what we have in our minds, and this means we will need to draw more than rectangles and arrows along the way.

I mean, we'll still draw rectangles and arrows, but with much more of a zoomed-in perspective.

They are also called algorithms. However, we won't draw the algorithms' software content; we'll draw from the users' perspective.

When talking about users, we're not only talking about the ones who download our apps. We're also talking about the app admins, moderators, editors, and whoever uses the app at any given time.

If your app has an admin functionality, that's also a thing to include in the user flow.

* * *

Drawing algorithms from the users' perspective is not rocket science. You just need to visualize every step of the way that your users can take. You should include everything that can be done in your app, not only the steps you want your users to follow.

Drawing the accidents is necessary too. For example, you should draw whatever needs to be done when a user quits the app while it is busy sorting photos.

By drawing, I'm still talking about rectangles and arrows, nothing more. So, if you want to draw a scenario, start with a rounded rectangle and an arrow. Then put a rectangle at the end of the arrow and write the action in it. Then draw another arrow and another rectangle. You can utilize UML diagrams, but it's not necessary in your case. You just need to make your decisions as straightforward as possible.

3- Spotting the Necessary Details

Yes, some details are necessary; some are not. However, if you don't have a grounding in development, you probably don't know what exactly is necessary. So, you include everything you can.

That's a good thing to do because the developers decide what's necessary to make the thing work; that's not your job. Your job is including whatever details you can think of.

Also, being objective when communicating the details is essential. You shouldn't say something like "users should be able to log in easily." In this case, "easily" makes it subjective. You shouldn't describe the feeling, as you did to designers, you should include the facts and definitions.

Using objective language in an analysis document matters. If you think you didn't make it clear enough at the first attempt, then try again. Write something more about the process. Define everything as clearly as you can.

* * *

Also, if you need help, you can get that from someone else. You can always utilize the professionals along the way to make everything easier and more straightforward for everyone.

After we have the software analysis document at hand, we need to know who can develop what we need. For that, you'll need to grade developers on your terms on what they can do for you.

Let's take a look at how we can grade developers.

How to Grade Developers

Grading developers is not an easy task, even for us developers and professionals. That's because there are so many parameters in the process, and every process is unique in itself.

I can't list every detail of the process in one chapter, nor even in one book; it'd take much more space than that. However, I'll share some tips with you on how to grade developers.

By grading, I mean grading with your project in your mind. It's not an objective evaluation and it doesn't matter who does it. So, we won't try to put developers somewhere; we just want to find the perfect fit.

The approach will be subjective, that's a fact. However, we can limit subjectivity in decision making, and to limit subjectivity, we need data. In the upcoming chapters, I'll give you a brief overview of the data needed.

Here are some tips and tricks on grading developers to work on your projects.

1- Objectivity

Your nephew may have coding skills, but those skills may not be relevant enough for building your product. So, you'd need to use a more objective framework for your thinking.

* * *

My approach to objectivity does not involve considering how well I know the person I'm grading. I may have known them for a while, or they may be a close friend or relative, that doesn't matter.

There will always be enough projects if you can succeed in the first few. So, you'll have room for the people you want to include in the future, no worries. However, we're trying to create a solid foundation here, so we need to grade objectively.

Trying to be objective on topics which concern people may be problematic sometimes. You should be objective about everyone, not just the people you know or don't know.

Another critical factor is circulation. There will be circulation on the development team if your project is successful. Being objective in your selection also brings you an opportunity on those occasions. If you write down your findings about the people you interviewed, you will be able to find out how they have improved over the years.

2- Productivity

In a way, productivity is overrated; you don't need to make every second of yours count on your projects. It may be better otherwise. That's also true for developers.

Developers are humans too; they need rest, spare time, time with family, etc. Even if your app should have been developed yesterday and you're still trying to find developers, they're still as human as you are.

Don't confuse productivity with working long hours. I have been working 12 to 16 hours per day for years, and I think you, as an entrepreneur, are probably used to a similar schedule, but that doesn't necessarily mean everyone you interview should work like that too.

We need to find a way to grade productivity. The first tip? It's not about limitless time. It's about how you use the limited amount of time you have. You can always have an option to put in the overtime, but the real achievement is getting there without sacrificing your personal life.

In development, more code doesn't necessarily mean better code. It may be true, or the opposite may be true, that's not relevant. So, you can't grade a developer by calculating the time they spend on any given task. It's the consistency of their work you should be looking at.

By consistency, I'm talking about keeping promises. It's one of the first things I cover when I teach development, and I can't emphasize it enough.

Producing the same results with the same input sets is essential, however, there aren't any identical input sets for development; We need to find a way to grade consistency.

Making promises and keeping them is not only an essential quality for developers, it applies to the rest of humanity too. If you can find someone with the confidence to make and deliver on challenging promises, then that quality is a definite plus.

3- Loyalty

Loyalty applies to both sides in a relationship, and the development of your first project is a two-sided relationship in progress.

That's because you have to trust someone to develop your ideas, and they have to trust you on keeping your promises. The promises you need to keep may include the business part of the project or paying on time.

If you want to work with people who keep their promises, you need to keep yours too. That creates synergy, and it's crucial for the early days of a startup.

You should also be transparent. Being transparent brings trust. First, make sure your approach is loyal and transparent, and you're someone who can be trusted, then it'll become easier to find someone with similar qualities.

* * *

4- *Quality Obsession*

It's a good thing, but it may be a bad thing too. Like water, it's the thing we can't live without, but even drinking too much water can harm our bodies. It's the same when it comes to quality obsession.

Quality obsession seems like another name for the term " perfectionism," but it's not quite like that. If perfectionism is the top of the mountain, we're talking about half the height of it here.

Being obsessed with software quality is important from the starting point of your project. Because when you start a project with low-quality code, it gets harder with time to build it.

Quality obsession and keeping promises form a perfect fit because they balance each other out. Making and keeping challenging promises without sacrificing software quality is crucial.

5- *Ability to Learn*

The ability to learn is another essential quality to have for a developer. Because we live in an ever-changing world, and technology is the flagship of that change.

Learning new things by yourself and keeping up with innovations becomes more and more critical. It would be best if you found someone with that ability to learn. If you can, you're halfway there before you start.

Because with someone who can learn, you can always improve yourself and your business.

Broad Comparison of the Programming Languages

Let's talk about the fun part; programming languages. They're the building blocks of any application from an AI to a website. They're everywhere; in almost anything you use nowadays, and they're not going anywhere.

* * *

So, we need to have some idea about them. I have listed some of them, and I'll take a look at each one from an entrepreneurship perspective. These are the languages I used to develop projects by myself or with a team, so I'll try to give you the perspective of what it's like to use each one without getting into technical details.

Programming languages can be grouped into any number of categories for any property they have, meaning I had to select one to group them in. I chose to group them as compiled and interpreted languages.

Compiled language means a compiler will generate machine code (zeros and ones) from the source code. They'll be much faster for compute-heavy tasks, but compilation takes time (with some exceptions).

Interpreted language means there will be an interpreter to read and execute your code step by step.

There are use cases for both of them, but without specific tasks at hand you can utilize pretty much any general-purpose programming language for almost any task.

So, let's dive in.

Compiled Languages

Java
Java is one of the most popular languages out there. It was released in 1995, and it's been widely used since then. You can use Java to code native Android applications, and it's probably where you'll see it most.

Other than the Android platform, Java runs on a virtual machine called JVM. JVM means Java can be used pretty much anywhere, for almost any purpose.

C/C++
C and C++ are low-level languages that can work anywhere, within

almost any condition. You can create servers, games, native Android or iOS apps, and so on with both C and C++. They're also fast. Really fast.

However, working with them is not that swift and simple. Since they're low-level languages, you need to manage every detail of the underlying machines to code them.

The compilation times are also not their most vital parts, so they're not widely used in non-performance-heavy tasks, like servers and games (thanks to Unreal Engine).

Go

Go is a relatively new language; it was created by the folks at Google in 2009. It was created to improve current compiled languages (especially C) in terms of compilation time, utilization of multiple cores in modern machines, ease of use, and feature set.

It's widely used in systems programming because of its speed and ease of use. You can create entire servers or small programs with Go, it's up to you. Go will offer you speed, ease of use, and convenience.

C#

C# (pronounced C-Sharp) is like Java but from Microsoft. They're almost identical, and C# is also widely used in almost any kind of application.

You can see C# in Windows applications and Games (thanks to Unity) nowadays. It's also widely used in cross-platform frameworks like Xamarin.

Swift

Swift is a language from Apple, built to replace Objective-C in the iOS programming world. It's widely used for native iOS development and it has one of the cleanest syntaxes.

Latterly, Swift is also being used in server-side programming. However, that's not widely adopted yet.

Kotlin

Kotlin is like Swift, but for Android. It was adopted by the Android platform to replace Java. They're interoperable, meaning that you can use both of them in the same app and everything will work fine.

Kotlin is widely used in native Android development. It also has some other applications, but they're not widely adopted yet.

Dart
Dart was developed by Google to create great client-side applications. It became popular with the Flutter framework, which is also a Google product.

With Flutter, you can create cross-platform apps with ease. It's getting widely adopted, and you may choose to work with it if your mobile applications don't use the latest device technologies.

Objective-C
Objective-C was the iPhone's official language when the iOS Software Development Kit (SDK) first appeared. Numerous developers and companies used it, and some are still using it because changing big application codebases can sometimes be challenging.

Apple kind of replaced Objective-C with Swift and they're improving Swift day by day, which means there will probably be no need for you to choose Objective-C over Swift.

Rust
Rust is a language used in creating high-performance applications. It's like a competitor of the language Go, and, like Go, it has its ups and downs. You can use Rust in server programming, where performance, efficiency, and concurrency are the primary concerns.

There are so many compiled languages that I have only included the most popular (in no particular order) to give you a brief idea about each of them.

Here are the interpreted languages:

Interpreted Languages

* * *

JavaScript

It's the language you see on the web every day, and it's currently the only programming language to work on all web browsers. It was created by Netscape, the popular web browser from my childhood, to make web pages interactive.

Since then, it's also been used to build server-side apps, cross-platform mobile apps, games, and a lot of other things in between. It's the language you go to when you can't decide what you want.

Python

Python is one of the most popular languages out there, thanks to machine learning and the artificial intelligence community. I believe its ease of use is the main reason behind its high adoption rate.

You can use Python for almost anything, but it's especially valuable in backend and artificial intelligence development.

It has a great community, and the collaboration that offers makes solving problems much easier. You can easily find someone with Python knowledge, however, grading them will be a challenge.

Ruby

Ruby is like Python's sibling. It's a great general-purpose language, which is used frequently in web development (thanks to the Ruby on Rails framework).

If you want fast prototyping and development on the web, Ruby can be your go-to language.

PHP

PHP is the third sibling. Web developers took to it with relish, and it's still used by a substantial community (thanks to the Laravel framework). You can start working with PHP quickly, and it's almost absurdly simple to prototype and develop some web pages with it.

It's also the language behind WordPress, so a significant majority of the web is built with PHP.

* * *

All of the above languages have their use cases. However, some of them overlap with each other. That's why you may need professional help to decide which one or which group of them to adopt.

Let's take a look at some conditions and scenarios surrounding their selection and use.

How to Select the Right Programming Language

When selecting a programming language, there are no clear boundaries between the right and wrong ones. Instead, we'll take a look at various scenarios and how we can act upon them.

We'll start with, you guessed it, the photo sorter app. Then we'll continue with a website where you can store your bookmarks. The third scenario will be a bitcoin ticker app, and the fourth one will be a hyper-casual game.

All of them are somewhat basic scenarios, and there's a reason for that: It's because when the process becomes complex it gets too far away from the generic so it's hard to analyze it in a brief section of a book. If you have more involved projects, you can always work with professionals along the way.

1- Photo Sorter App

Here, we'll have it on both iPhone and Android phones. We'll have two options on the app side: First, we can work with an iOS developer and an Android developer. Second, we'll use a cross-platform framework.

If we go native, we'll have to find an iOS developer who can do several things simultaneously. These will start with creating photo sorting algorithms. After that, they'll need to develop the app from start to finish, meaning implementing its design and functionality to your specifications.

We also need them to have experience in making the app go live. It may be a challenging process, so they'll need to have at least one app in the App Store.

It'd be perfect if they knew a thing or two about Android development too, that's because developers are better within communication with each other. It's like talking the same language. There's no need for them to be an expert though, it's solely that it would be nice if they possess some knowledge.

Then, we'll need an Android developer. The same qualities are needed for them too.

We can always find someone who can develop both iOS and Android apps by themselves, but it doesn't necessarily reduce development costs. So, we need to think about our unique conditions.

Another option is working with a Flutter or React Native developer to create a cross-platform app. The photo sorter app will be a challenge for them, but they can develop the app using both without losing any performance whatsoever in terms of the feature set.

2- Bookmarks Website

The second scenario is a website where we'll keep our bookmarks, and we'll need at least one backend developer for that job. After that, we'll need to decide if we want it to have a separate frontend or not.

Let me explain:

There are excellent frameworks out there, especially on the web, which make it easy for developers to build something, and you should be using one of them in your project.

The most popular 3 are Django, Ruby on Rails, and Laravel. Any one of them can be a perfect fit for your project since they run websites from the smallest to the biggest.

For Django, you'll need a Python developer. Python is one of the most used languages out there, so there's a good chance that you'll find a ton of people who know Python. The thing is, they don't necessarily know Django, so you should ask for some reference points and a portfolio before starting work.

However, the same thing may not apply to Ruby on Rails and Laravel. They both are one of the biggest reasons why the language behind them is so popular.

Ruby on Rails gives you the flexibility and ease of development, although getting started is a bit harder than with the rest.

Laravel not only has those advantages too, but also the ease of getting started.

After we decide who to work with on the backend, we'd need to decide whether we'll utilize frontend frameworks or not. If we do, there should be at least one other developer who can code frontend apps.

They'll be JavaScript developers. However, as there are countless frameworks out there, it's better to wait for a bit. We'll return to this topic in later chapters.

3- Bitcoin Ticker

If you want to have a bitcoin ticker app, you'll need two parts: First, the API element. API stands for Application Programming Interface, which means the backend here. Second, is mobile development.

In mobile development, the scenario is more or less the same as that for the photo sorter app, meaning we'll need at least one mobile developer, but there may be more.

We can utilize the frameworks from the bookmarks website part in the backend phase. They all have API serving options to use with client-side development. Also, if we need high performance, we can use Go or Rust to program the backend part.

I put the bitcoin ticker here because I wanted to show you there are no clear boundaries between any of the development phases. Deciding what to use is more of an art than a science because you can use any of those tools to create the same things.

The main point here is what you have around you. Your unique conditions determine what you should choose and where you should go with them. So, to start with, and as a priority before everything else, know your exact conditions at hand.

4- Hyper-Casual Game

Here's a fun project. However, the development phase wouldn't be that much fun. We need at least one game developer using a game engine like Unity or Unreal Engine.

Then we'll need to have at least one artist. They may be 2D or 3D artists; it depends on what kind of game we're trying to make.

Then we'll need a crystal-clear game design document, which should include all the parts and mechanics of the game. Game development can be fun for experienced developers, but you'll see the pain in the eyes of newbies.

If we go with Unity, we'll utilize C# language. If we go with Unreal Engine, we'll go with C++. There are more engines out there, but these two are the most popular ones.

Let's look at the languages, tools, and frameworks for web & mobile development. Then we'll take a look at what the "full-stack developer" is and what they can do for you.

Languages & Frameworks for Web Development

There are so many popular programming languages and frameworks to create web applications that I won't list all of them. These are some of the most popular languages and frameworks, but there are more.

* * *

Python

Python is an excellent language for beginners, as well as seasoned experts. That's why there'll be a plethora of Python developers, and you should be aware, there are beginners and experts. So, not everyone who knows Python can do everything with it.

Here are the two most used web frameworks, written in Python:

Django

Their official website tells you, "Django is a high-level Python Web framework that encourages rapid development and clean, pragmatic design. Built by experienced developers, it takes care of much of the hassle of Web development, so you can focus on writing your app without needing to reinvent the wheel. It's free and open source."

So, it has basically whatever you need to develop for the web. However, it's like almost every high-level web framework, as each framework of this kind also offers that kind of infrastructure. So, it's not the benchmarks that will decide which one to use; it's you and your fellow developers.

You may know some of the companies that use Django: Instagram, Spotify, The Washington Post, Pinterest, and Dropbox. I don't know which ones are current users, but a swift bit of research will show you that they have used Django at least once in their lifetime.

Flask

Flask is Django's baby sister. It's a "micro" framework.

They explain micro like this: "Micro does not mean that your whole web application has to fit into a single Python file (although it certainly can), nor does it mean that Flask lacks in functionality. The "micro" in microframework means Flask aims to keep the core simple but extensible. Flask won't make many decisions for you, such as what database to use. Those decisions that it does make, such as what templating engine to use, are easy to change. Everything else is up to you so that Flask can be everything you need and nothing you don't."

As you see, Flask doesn't offer everything from scratch. It utilizes extendibility to create your structure with your tools.

In some cases, it's more valuable than the opposite. However, the decision here is highly technical, and it's heavily dependent on your fellow developers' knowledge base.

Here are some of the companies that use Flask: Netflix, Uber, Patreon, Airbnb, Samsung, and Reddit. They may not use Flask everywhere, but in things like microservices, Flask makes everything easier.

Ruby

Ruby is a language built for ease and the delight of development. Although it has applications other than the web, they're not that popular. So, if you find a Ruby developer, they probably know one of the following frameworks:

Ruby on Rails

As they stated: "Ruby on Rails is not a minimalist framework, it's a metropolis. One filled with all the major institutions needed to run a large, sprawling application"

Ruby on Rails tries to include almost every tool you'll need to build a web application. And it is exceptionally good at including them. Once you get used to it, Ruby on Rails is clearly easier and faster than any other framework. By faster, I mean faster development, not faster performance.

Companies using Ruby on Rails include Shopify, Github, Basecamp, Airbnb, Kickstarter, Hulu, etc.

Sinatra

Like Flask is for Django, Sinatra is Ruby on Rails' baby sister. They describe Sinatra, not as a framework, but as a Domain-Specific Language. It's used to create web applications quickly and with minimal effort.

However, you can't do everything you do on Rails, with Sinatra, at least not from the start.

Companies using Sinatra include Apple, Heroku, Github, the BBC, LinkedIn, etc.

PHP

PHP is a simple language with which you can do great things. Because of WordPress and Laravel it's here to stay, so let's take a look at two of its frameworks:

Laravel

According to the Laravel website: "Laravel is a web application framework with an expressive, elegant syntax. A web framework provides a structure and starting point for creating your application, allowing you to focus on creating something amazing while we sweat the details."

Laravel gives you the starting point for any web application. It's not necessarily easier or harder than Django or Rails, it's just a choice to be made.

Companies that used have Laravel at least once include 9GAG, Pfizer, and the BBC.

Lumen

It's, you guessed it, Laravel's little sister. It's also made by Laravel's developers, and it has faster request handling than Laravel. It is a micro-framework, which means it's suitable for small tasks and extendable for bigger ones.

JavaScript

JavaScript is everywhere. You can use JavaScript in the backend, in the frontend, or a mobile app. We'll look at some of the most popular JavaScript frameworks for backend and frontend development.

Backend: NodeJS

NodeJS is not a framework, it's the thing when you write JavaScript on the backend side, which means you'll hear its name frequently if you're talking about JavaScript in backend development. You'll also hear about something called Deno, which is one of Node's successor candidates.

Backend: ExpressJS

Express is the go-to framework when you use JavaScript on the backend side.

Its website says: "Express is a minimal and flexible Node.js web application framework that provides a robust set of features for web and mobile applications."

It can be used as a starting point for building NodeJS apps, and it's highly extendable.

Frontend: VueJS & ReactJS

In Frontend, you can use JavaScript as well. There are many front-end frameworks out there and you can use any of them to do almost anything you want.

It's not a discussion about performance or security, it's about preferences. You can use any of them, pick some from the most popular two, or develop with older ones like JQuery.

Selecting a frontend framework is more like a hiring decision than a technical one. There are plenty of developers who use Vue or React, and you can choose which ones to work with.

Others

As I said earlier, there are countless languages and frameworks which are widely used in web development. These include, but are not limited to, Spring (written in Java), .Net (written in C#), Angular (written in JavaScript), Martini & Gin (both written in Go), and so on.

Language and framework choice is critical at the start because you'll need to design the business aspects to play along with it, like hiring, extending, and employee circulation.

* * *

Let's take a look at the languages used in mobile development.

Languages & Frameworks for Mobile Development

There are plenty of alternatives to choose from when selecting a language and framework in web development. However, in mobile development, that's not necessarily the case.

At the time of writing, there are two major platforms: iOS from Apple and Android from Google. These are the operating systems that are running on devices.

Apple doesn't allow the use of iOS on non-Apple machines, so the devices that run iOS are made by Apple. However, that's not the case for Android, because Android is an open-source operating system, and anyone can use it on any device.

There are some standards within each operating system, and there are also cross-platform languages and development tools. Let's look at what we can use in mobile development by platform.

iOS Development

In iOS development, we used the language Objective-C from the App Store's early days. Apple released Swift in 2014, and Swift became the norm in iOS development after its release.

Objective-C is a C-based language with a C-like syntax. It has the typical features that a C-like language can offer. Swift is more like a modern language. It's simpler and beginner-friendly. Apple has been working on making Swift a better language for years, and they're pretty good at making things better.

We must have a Mac to develop iOS apps. There's no other official way to do it. Within the Mac, we'll use a program called XCode, also made by Apple. XCode is the Integrated Development Environment (IDE) for iOS, macOS, tvOS, and watchOS development.

* * *

With a developer account from Apple and a Mac at your fingertips, you can make any app available to any Apple App store.

You need a developer account to sign and send your apps to the App Stores. You can get an Apple developer account for yourself or your company by signing in to the Apple Developer website (developer.apple.com) and creating one. You'll need a membership to allow you to send your apps, though, and it'll cost $99 a year. At least, it does at the time of writing.

After setting up your developer account, you can use it on your fellow developers' machines as well. Then they'll be able to sign the apps with your account details and send them to your App Store account.

Android Development

In Android development, you don't need to have any specific device. You just need to have a developer account on the platform where you want to send your app. At the time of writing, it costs a one-time payment of $25 to have a developer account on the Google Play Store.

The lower barrier of entry is one of the main reasons why there are more Android developers than iOS developers.

We just need a computer with an operating system on it, no matter whether it's Windows, or macOS, or Linux. We use a program called Android Studio to develop on the Android platform.

Android Studio is a great IDE too, and with it you can do pretty much everything you need to do for the Android platform.

The two main languages used for Android development are Java and Kotlin.

Java was the first official language of the Android operating system. It is a concrete language and one of the most used globally, so it was a good starting point. However, Java is not necessarily a beginner-friendly language. So, another one was needed.

Kotlin was first designed in 2011 and version 1.0 debuted in 2016. In 2017, Google recognized Kotlin as an official development language for Android alongside Java and C++. You know about C++, it's the lower-level language to use in both platforms.

We can also use lower-level languages to develop apps for both the above platforms. And because they're running apps that are compiled to machine code (ones and zeros), there's a chance for us to utilize this mechanism to compile from other languages.

That brings us to the cross-platform frameworks.

Cross-Platform Development

First of all, let me clarify: Cross-platform development is not native development. The frameworks may be utilizing the platforms' components and styles, but they're not native apps.

The main things that make a difference here are the platforms' specific features, needs, and operating systems. For example, Apple releases new iPhone models with brand new features alongside the new iOS versions. They're highly optimized to work together in a small machine, and they're good at doing what they are designed to do.

However, there's a need for cross-platform development, and there are some critical reasons why: The first one is the budget. Not every person company has the budget to form different teams for different platforms. Not every developer necessarily knows the specifics of all the devices available.

So, there is cross-platform development. It's putting a layer in front of the operating system and working on that layer. It theoretically means everything will be fine with just a little more processing power.

If you have a small or moderate app that doesn't need native performance, the modern cross-platform frameworks can give you native-like benchmarks. At the time of writing, I'm getting almost native-like performance from Flutter, in a big enough app.

Using cross-platform tools is cheaper. It's also faster to develop because you don't need to develop the same app again for different platforms. You just need to decide if you're OK with the native-like performance instead of a real native one.

There are many benchmarks all around the internet, and you can find some to inform you about your particular needs. Or you can always ask professionals in the field.

Cross-platform development is quite popular these days, so there are lots of languages and frameworks to choose from. I've seen almost all of them in the wild, and I can suggest working with only two of them: Flutter and React Native.

Flutter, like Android, is developed and maintained by Google, and it's the best cross-platform framework I've seen in years. You use Dart language and Android Studio to develop apps with Flutter. Note, that even if you can develop with it, you must still have a Mac to build and send your apps to the Apple App Store.

React Native is developed and maintained by Facebook, and it's widely used because of the underlying language, JavaScript. Almost every modern developer has probably seen or worked with JavaScript at least once, so it's probably a safe choice when you don't have many options.

Like cross-platform frameworks, there are cross-platform developers, who we call, "full-stack developers." Let's look at who the full-stack developers are.

What is Full-Stack Development?

Full-stack development is the term used to describe developers who are capable of developing both on frontend and backend. However, it's not a strict term, and it doesn't contain any specific technology and framework by itself.

"Frontend" doesn't necessarily mean what you see on a web page. It can also be a Windows or macOS application, a mobile app, or anything like that. Frontend development is necessary for creating the user-facing parts of the applications.

The tech stack mainly consists of HTML, CSS, and JavaScript, if we're talking about a web frontend. It can also be any technology in the last part of mobile development; it's not necessarily telling you what the developer does.

This is also the case on the backend side. Backend development means the development of the non-user-facing parts. Like the servers, Application Programming Interfaces (APIs), automation, and so on.

The term full-stack developer describes a person who can develop with at least one server-side (backend) technology and one client-side (frontend) technology. They don't necessarily know all of the available technologies, and they probably don't.

You can work with a full-stack developer from the prototyping phase to developing the main application. They are useful every step of the way, so you can work with them with ease if your application only consists of a frontend and a backend.

Here comes the critical point: There are more parts in the development process. There are parts like infrastructure on the backend and interface animation on the frontend. Not everyone who can develop a full-stack can also understand the details of those elements.

This means you should select who to work with wisely. The recruitment rules from previous chapters also apply here.

Let me tell you my story.

My entrepreneurship journey started when I was eight. I was providing a taxi service to my classmates. Not a regular taxi service; it was more like "walk with me and pay me for that" kind of taxi service.

Anyway, I believe my entrepreneurship journey started there.

My development journey, however, started with a language called PERL when I was 12. There was no internet for my personal use at that time, so I learned it from books. I developed my first program, a video game saloon operator, with PERL.

When I was 14, the internet had arrived, but I didn't know how to acquire information from it. So, when my father gave me a book on a language called Delphi, I started to work on that and coded a math practice tool. I was preparing for an exam back then and I used it to practice my speed calculation skills.

Then I stopped for a while. For around four years, I did almost no programming of any kind, instead, I started working on design.

I went to university, but decided that I didn't want to be there, be graduating in economics, or just be a designer. So, I went back to coding. I used my entrepreneurial skills to develop some ideas and worked on them. This time, I was aware of the internet, and I was aware of learning from it via tutorials.

I started again, but with the PHP language this time. I learned to code while developing my projects, then I started to develop for clients as a freelancer.

Merging entrepreneurship and development skills enabled me to work and live wherever I wanted, and I made use of that benefit. I would call someone that I knew needed a website and offer to develop it. If they said yes, I moved nearby, made some friends, and stayed with them.

After that "web development with pure PHP" period, I discovered frameworks. Especially the Drupal framework, which was a game-changer for me. Drupal was (I believe it still is) a highly customizable and extendable website building tool and framework that could build almost any kind of website. I utilized it big time.

I started to develop websites in a day, and my resources began to increase accordingly. I used those resources to learn more about what I was doing and what was possible. Another turning point in my career came next, with a phone call.

After that call, I started to work full-time for a company as a developer for the first time in my life. That is how my professional career kicked off.

I learned to develop mobile apps there and started on my side projects. One of my side projects, an app called Dietitian, reached the top of the App Store chart for almost a week in Turkey. That made me realize I should think of getting back into the entrepreneurial side of things.

I got back to entrepreneurship by forming my own software development company when I was twenty-four. From then on, I started to work on my business. After the period of developing for clients, I realized that I coded with numerous languages and tools.

I started to work as a consultant, as a side project. I acquired some clients, and that went pretty well. I set up more businesses and did further consultancy work, while learning extensively about my craft.

That's how I became a full-stack developer: By using almost every modern language and creating something with each of them. It has been an incredible journey so far, and I'm looking forward to it continuing every day.

And that's also why I'm not only a developer, but also a consultant. Because I wanted to learn all the ups and downs of the software development industry, and I still do. That created a field in which I can work on projects with people I believe in.

And that may be the reason why we could work together sometime on your project too, who knows?

Now, it's time for the third chapter: Negotiation.

You may already have negotiation skills in place, but you should still know how to approach developers and work with them. Let's take a look at how exciting negotiating with developers can be.

CHAPTER THREE

Negotiation

How to Negotiate with Developers

Negotiating with developers can be tough. I've been on both sides of the table many times, and I want to share some tips with you.

But first, let me explain what I mean by negotiation:

Let's say you found the developer you want to work with. The next thing to do would be to make them an offer and wait for them to accept, then everything will be fine, right? Well, it's not always like that.

There is a thing called the ego, which, while being a good thing, is also a bad thing. You need to know what to do before and during the negotiation, and your ego, and that of your candidate will play their parts.

The process may be different for everyone, however, there are some constants that you should be aware of. Here they are:

1- Know What You Have

First of all, you should know what you have available. What resources you have, like how much money you can pay, how much time you can spend with them, what you need to arrange before working with a developer, and so on.

The budget you have does make a difference, however, you don't need to have a great deal of money to begin with.

Time is a critical thing, because they'll give you their time, and you should reciprocate. Balance is essential when it comes to working on a project. Also, spending your time on it brings motivation by creating value.

Don't underestimate your power to motivate people. It's also your power to convince people to do things they wouldn't do otherwise. That's a hard thing to do, and having something to offer is a powerful tool.

Know what you have and know your limits. Your limits make you who you are. In an honest entrepreneur to developer relationship, everyone should know their limits and strive to extend them continuously.

2- Know What You Need

Knowing what you need in your business is about understanding your unique conditions. Also, there is a need for you to find out what your exact development needs are. That may sometimes be trying, so a little outside help could ease the pain.

Because of the difference in perspectives, you might not be able to see all of the needs in terms of development and there is a small problem here:

If you talk to a developer who's eager to work with you and tell them about your unique situation and your project, they'll approach your project and what they think you need from their own perspective.

Of course, that may not be the best perspective for you. It won't hurt you in the beginning, but you might want to nail down the technology choices, the experience of the person you're talking to, etc.

* * *

Or, if you talk to a great developer who's not interested in your project, they might give you some advice anyway, because they know their craft and are willing to share. However, this also may not be the best advice for you because you may not be able to find someone to develop your project in those terms.

3- Know What They Have

Then, you need to make sure you know what they can offer. It's the thing we discussed in previous chapters. You'll want to know the capabilities of the people you're working with.

Because you'll be a startup, and being a startup is arduous by its very nature, you'll want to work with people who make the process easier, not harder. And let me tell you a secret, making a process more comfortable is not something everyone can do. Because it's tough.

You should find out about their skills and how aligned you are with them. That's the best way of starting to develop a startup project. The more you are aligned with your team, the better work you'll produce together.

Working together in a startup is like marriage. You'll spend endless time with your team, and knowing what capabilities they have will strengthen your hand when the time comes to face the outside world.

That's the main point here. You'll be out in the big, bad world selling the things you and your team built, and your users will want countless things from you and your team. You should be firm in saying "yes" or "no" to customer requests. It means you know what your team can and can't develop, and in what kind of timeframe.

4- Know What They Can Offer

Let me reveal another secret: Developers have limits, just like you do. There is no black magic; it's all code. Also, not everyone can develop everything.

It's typical of developers to underestimate the projects they have.

That's one of the things we'll cover in later chapters in detail, but let me quickly explain:

Think about a situation where someone wants you to estimate how long it would take to climb a mountain. If you climbed that mountain, or a bigger mountain, you would know what the challenges are and how to face them. Your estimations about climbing that mountain would be more or less accurate.

However, if you haven't climbed a mountain of that size, but only smaller ones, and you really want to climb that mountain that may cause a problem. You may not have a chance to climb a bigger mountain in the future unless you try to estimate as accurately as you can. It may be an approximation of accuracy, but there are things to note here:

First, it's essential to find someone who can estimate without too much guesswork, but that can be difficult. You may not always be able to find someone who knows everything they could.

Here's the thing: Estimations are estimations. They're not facts, and they probably won't work as planned. It'd be best if you accept the fact that your fellow developers may not precisely know how to do everything that's needed. It's about you believing in them and motivating them, and them doing their bit by showing the ability and willingness to learn.

Let's take a look at different approaches to working with developers. Someday you'll have to pay them, so it's best if you are aware of the different kinds of things you might pay for.

Time-based Approach

First, we'll look at the time-based approach. It's simply paying for development time. We're working with a freelance developer, leasing their time and effort to develop what we want.

It is a system that works throughout the world every day. But there are conditions in which it works and those in which it doesn't. Let's check out those conditions:

1- You Know Exactly What to Expect

You've done your idea, business, and software analysis. You've created the necessary documents and know exactly what to expect. Even in that case, you may not be aware of some of the details of the work.

Freelance work is about spending time on tasks and reporting back. So, if your tasks aren't straightforward or have some gaps, it costs you money. Everything may look fine at first glance, but after starting work, things may seem to slow down.

In those situations, you should re-analyze both your work and who you work with. Knowing exactly what to expect is one of the critical points of working with freelance developers.

2- They Know Exactly What They're Doing

You also assume that they know what they're doing and are willing to do it for you. You may think it's only about money, but you could be wrong. Like any other job out there, developers work better when they are motivated, and development motivation comes from two things: Expectation and knowledge.

The bigger the expectations of a developer, the bigger their motivation. The more they know also falls under the same principle. Just like any other group of people, not all developers are the same. Style over substance makes some freelancers look better than they really are. So, it's easy to be sold on them if their marketing is persuasive.

You should know something about your fellow developers' capabilities. As we discussed in the previous chapters, you're responsible for choosing the right people for the right job.

I have seen some real-life problems like these: An entrepreneur once started working with a freelance developer. The developer worked on the project for two months, showed some progress, but the work was not done.

When I arrived on the project, and after talking with both parties, I checked the code which had been written - we'll look at how I did that in the upcoming chapters about where to store the codes. I immediately realized that there was no more than a week or so's worth of work there. The developer said that he had been working on the project, but the reality was different.

The problem was not the developer's work, which was good, and, as he said, he had almost finished the work. However, the problem was in his work ethic, in that he reported the process was taking longer than it actually was. He built the app in a week, but extended the timeframe to get more paychecks.

Of course, that is a particularly bad scenario, and it doesn't necessarily mean you'll see that kind of thing. However, these things exist, and you should be aware of them.

3- You Defined the Tasks Clearly

Even if you know exactly what to expect, that doesn't necessarily mean that you wrote everything down clearly. That's especially important in freelance work because most of the time, they're on their own, developing your app in their conditions.

And their conditions are not necessarily the same as yours, so they can't think like you. Details make all the difference in the execution. Having a broad idea may suggest things to people, but they may well differ from your vision.

Therefore, your task is to define the developer's tasks clearly. I know it sounds like a big ask, and that's why I'm helping entrepreneurs and businesses with it. Defining precise tasks is one of the most challenging and rewarding parts of the process, so do your best.

4- Everyone Acts Ethically

As you saw in the second point, not every person acts ethically. It's a base expectation, and while it should be there every time, sometimes it isn't.

That's a hard problem to solve, but there is a solution. If you know what to expect and give clear directions, then there'll be no way to act unethically. It's like having an agreement with developers to work under such conditions without having it put down in black-and-white.

Of course, you need to have a written agreement, but I'm talking about something else here. It's more like closing down all the options but one. Making people act ethically is challenging, but that doesn't mean it's impossible. Even if it looks like it is, as an entrepreneur you have the power to make the impossible possible.

I have also seen unethical entrepreneurs too. Some people don't pay on time, for example, which is always a bad idea. Please be ethical and do your best to keep your promises. It helps to make the world a better place.

5- You Have a Backup Plan

After you've defined the task clearly, it's time for you to have a backup plan. The backup plan is about working with someone else in case of problems. The main point here is preparing yourself and your business for the worst possible scenario.

Preparing for the worst is not calling for the worst, so don't underestimate plan B's power while doing everything in yours to make sure plan A works. It makes your process stable, and you'll be good to go if need be.

6- You Have a Control Mechanism

It would be best if you had a continuous control mechanism. Being a freelancer doesn't necessarily mean that they do whatever they want and work under whatever conditions you want. You should have mechanisms in place to control whether they're working on your project or not.

* * *

You can succeed in working with freelancers under those conditions. If you make sure the conditions are as flawless and error-free as possible, you're good to go. It'll always be better if you have plan A and plan B when working with freelancers because you should always expect the unexpected.

There are also different approaches to getting the job done, and one of them is the project-based approach, which we'll take a look at in the next section.

Project-based Approach

Like the time-based approach, the project-based approach is also a way to measure and pay for the work. What makes the project-based approach unique is that you set a fixed project cost which is the same no matter how long it takes to finish the project.

It's an excellent way to work, especially if you're in a position where you have some money but don't want to share equity, and the developers have enough expertise but perhaps don't wholly believe in the project.

Let's face it, there will be times when we can't force developers to like our project, and that's completely OK. They may also have realistic and logical reasons why they form that opinion, and it makes a difference.

I have seen developers hiking their prices when faced with a project they didn't fancy. And, in the opposite scenario, I have experience of developers who were happy to work almost for free, just to be involved in getting the first of a product into the marketplace.

The success of the project-based approach depends on these conditions:

1- You Know What You Want

That should be obvious, but it's the number one problem amongst entrepreneurs in the software development field. If you don't know what you want from the process, you can't get it no matter what. The first step is knowing exactly what you want.

Let me give you an example: I worked with an entrepreneur who was struggling to get his app on the market after the developers had finished with it because it was incomplete. I took a look at the codes and the app itself and realized it was almost half-done.

The reason was that he didn't know what he wanted. When I took a look at the documents he gave to the developer, I realized they were also half-done. So, the thing that he wanted and the thing he told the developer that he wanted were different animals.

Then there were other problems, like finding developers who are willing to work on a half-finished project. Believe me, it's harder than you think. Half of the project is worse than none because the developers have to tackle another learning curve.

It's better to be the first developer than the second one. The second one can get it to make sense, but being the second one is the worst. You know there was a problem once, and you know they'll want you to fix a problem you didn't create.

From a developer's perspective, it's best to know what you want and tell them all the necessary information they need to know.

2- They Want to Work With You

In the first developer and second developer problem, we glimpsed reasons why they perhaps wouldn't want to work with you, and there may be many other reasons too.

It seems obvious when you think about it, but it's not always the case. Let me explain it in more detail:

In one project, I faced five developers working on the same codebase. The project went well throughout the entire first phase, but then something changed, and it started to slow down.

It's not unusual for software development projects to start to slow down after a while, but that case was different. They couldn't find out why it slowed down because it wasn't that obvious.

It took some time for me too. Then I realized there was one person making mistakes who was responsible for the infrastructure (which is called dev-ops, we'll have a look later on), and continuous integration. Those mistakes caused builds to be slower than they had been.

After checking in and working with him, I realized that he didn't want to work there. Initially, he needed the money, and then he continued with the work even though he didn't believe in the project., He didn't like it at all, and he was the only one on the team who was being paid in line with the time-based approach.

So, make sure your team wants to work with you on your project.

3- You Defined the Project Clearly

If you don't define everything from start to finish, that's understandable. If you don't work with a time-based approach, you don't need to pass on every single detail to make it work because someone is willing to take the risk with you.

The main risk here is that you may not be aware of every exact detail. Consequently, it may take more time than expected. If you work project-based, the developer must cover the difference, so you don't need to worry about it that much.

The thing you should think about is defining the project's boundaries well. It's a good idea to write down what you want from a user's perspective and make it highly detailed. With a straightforward design and a clear document, you're good to go.

4- Everyone Knows Their Limits

And the last one: Developers are human beings and they have

limits. Interestingly, you have limits too, for the same reason.

There is a need for everyone to be aware of their limits.

You may not have sufficient technical knowledge to define your project clearly. Or you may not have the necessary budget to build such a big project. If so, then increasing your limits is essential, but first, you need to know them.

Likewise, the developers who don't know whether they can code your project or not shouldn't accept your splendid offer, even if it's a tough decision.

Because of our limits, we are creative. Creativity is a vital part of the entrepreneurship journey, so don't worry about being limited, just try hard to be creative inside your boundaries. They'll increase in size.

It's also true when it comes to developers. They may not have worked on a project like yours, but a willingness to learn may give them the courage to work with you. The main thing you need to calculate is the capabilities of the developer.

No matter how willing they are, there may be some subsequent problems which you could do without. So, grading developers, as previously mentioned, is essential before working with them.

Sharing Equity & Vesting

Sometimes, you may not have enough money to work with a developer or a development team, but there are alternative methods you can use to make your project go live.

That's not the only reason why sharing equity exists. You may (and perhaps should) want to share some portion of the company with those who helped to build it. It's crucial for you to understand the logic behind sharing equity and vesting.

By sharing equity, I don't mean giving away more than half of your company. If you're the entrepreneur and you're responsible for who gets what, always keep more than half of the company in your hands. Of course, if more people are involved, it becomes harder and harder.

Also, there is vesting, which means receiving an agreed portion of the shares after an agreed amount of time. It works like this: You and your co-founders will have an equal number of shares. However, you may want to make sure that everyone is on board and working for the same company goals. So, you include a rule which states that no one can receive any shares until they have worked for a year. And, after that year, you may get all of your shares in, say, four years.

That's way better when it comes to working with someone who you haven't worked with before, especially if they're your friends. Methods like vesting can keep your friendships alive alongside your business, please consider them.

A common problem for entrepreneurs is deciding what portion of the company to share. Let me give you a framework on that subject:

If you have co-founders, try to calculate their value by putting what you knew about them out of the equation and thinking about their value to the business. They may be doing the fundraising, and because you might be lost without fundraising, that makes them crucial to your business. Then, in the same exercise, look at yourself as honestly as you can. Having done that, you can decide who's worth what portion of the company.

Some people may be essential to get a company up and running; however, consistent work is needed to qualify for a significant share in the business. You wouldn't want to pay unlimited money for a little work. If anyone's involvement is limited, consider paying them instead of giving them equity.

Because the company will demand continuous work from all of the co-founders, and you might not want to give half of it to someone who's not there, vesting is also a way of protecting the business.

After you have reached agreement with your co-founders, it's time to think about sharing equity with developers, designers, marketers, and so on. Let me give you a different framework for that:

* * *

Think about the money you can spend. If it's more than you need to spend on development, that's good; we'll get back to that. If not, think about the projected value of your company, and try to be extremely honest while not being hard on yourself. Your company may never be worth billions of dollars, but you should already know that.

Determine the projected value of your company by making some financial projections. Once you start, it's not that complicated, but you do need to consider your company's future. After you determine the projected value, you can start thinking about what to offer to developers and other employees.

After projecting your company's value for the future, it's time to calculate the portion you should give to developers. There is no hard and fast equation here, but anywhere in between five to twenty percent works for those who make it happen.

That's because you don't have enough ready money to pay out. If you do, and you should, things are a bit different. If you're paying a salary to those who built the company from the ground up, you should familiarize yourself with an "Employee Option Pool."

Setting a pool of around ten percent of the company is standard for tech companies. If ten percent of the company sounds huge, don't worry. It is shared among many people, not just one.

Here's a scenario where you can calculate how much to pay from the Employee Option Pool:

If you agreed to employ a developer on a full-time salary, then you'll have the option not to give them any shares at all. The trouble is, that will not motivate them, and you can deliver motivation with some shares.

It's best to give 2 to 5 percent of the company to the first employee with a five-year vesting period. After the first employee, you can start to decrease the amount so that the company won't run out of shares.

Two percent of the company may seem too low initially, but the main problem is double-edged. It may seem low to one person, yet too high to someone else. It's about projections. You may be the best person on earth to make financial projections, but these are still projections; they're not real yet.

You will encounter these contradictions from time to time, but let's keep focused and calculate as accurately as possible. While doing that, you should also set some standard procedure about sharing equity. You'll build that eventually, but thinking about it keeps your shares in your hands from the start.

Another scenario:

Say you agreed to pay full-time money to an outsourcing development team's staff. In this scenario, you don't have to part with any shares. Because you don't give shares away for the performance of work, you do so for loyalty. Since an outsourcing company can't stay with your company, you may well choose not to grant them any shares.

And another one:

Say you are working with a developer who is capable of being your future CTO and working with you for a fee which they could easily get elsewhere. In situations like this, consider them as being your co-founder, because they may create more value that way. The equity sharing equation will also change, since they're not just a regular employee anymore but also a co-founder.

We've discussed sharing your company; let's discuss sharing your ideas.

How to Prevent Idea Stealing

The stealing of ideas is a fact of life in the entrepreneurial World. It's hard to deal with it, and no matter how protected your idea is, there is always a chance that someone will steal it. After all, you're trying to showcase it to the world someday, then it'll be accessible to everyone.

* * *

However, there are some things to know about ideas that make stealing them hard. Let's talk about them:

1- Patents

You can protect your idea's details by patenting them. That's a way to say to the world that you made the thing, and no one has a right to make it again. It seems an excellent way to protect hardware ideas, however, in the software business, patents may not protect you that much.

I'm not a lawyer, so I can't tell you what patents to get or under what exact conditions they protect you in the country where you register them. You should consult a lawyer for professional advice about this topic.

I'm just aware of the fact that multiple people can have the same idea at the same time, so patenting a broad idea like an "automated taxi service" is problematic.

2- Non-Disclosure Agreements

There's also another way to protect your ideas from your team and any people who you showed them to. It's called the Non-Disclosure Agreement (NDA). It ensures that what you share will be protected by the person you share it with.

You might be showing investors a business plan, or showing wireframes to your designers and developers. It'll protect you within a small circle of people you can trust.

Again, I'm not a lawyer, so I can't guide you through the exact details of such an agreement. I consult my lawyers on writing these documents, and you should too.

There are other agreements that may include an NDA, such as employment contracts, so you may not need to draft another one. And again, consult your lawyer before entering into any agreements. They're there to protect you.

* * *

3- Finer Details

By the way, there are things you can do to protect yourself that don't include any agreements, and I can help you with them.

First of all, ideas are living things. They grow over time, and they get better and better with finer details. At least they are meant to be that way. So, if someone steals your idea, the chances are that they may create an entirely different thing than you imagined.

That's because the devil is in the detail. Everyone sees the world differently, and that's also why you should be aware of your unique conditions. Everyone would handle the same idea differently. Like making music, give five producers the same sample and they'll create five completely different tracks out of it.

Think hard about the details, and share them with your inner circle. There will be some people who must know everything about your ideas, and you have to trust them implicitly. Pick your inner circle wisely.

4- Execution

You'll know that Facebook's story began with a non-original idea, like almost all startups and companies. There may be original ideas in the process, but there are no original ideas in general.

Ideas are generated with experience and by seeing the world in one's own way. We're all living on the same planet, but we're experiencing different things than each other, and that's what makes everyone unique in some way.

It may make you unique, but that doesn't make your ideas unique. If you've been educated in some fashion, that means you were given a world view by someone else, at least once, anyway. It is called influencing. People have a remarkable ability to influence others to do their bidding.

* * *

It creates environments for people to live in and people living in the same environments tend to think alike. That's why many people can have the same idea simultaneously.

What matters most here is the execution of the idea. Take Facebook, for example: It wasn't the first social network. There may have been hundreds of social networks built before Facebook, but what was the point that made Facebook so successful with the same idea that hundreds failed with?

It's the execution. Execution is the way you use your particular touch in your unique conditions to make things and events happen. It's the way you do something, and it's what makes your detailed ideas unique too.

If you take a great idea and execute it poorly, then you have written your very own fail scenario. Or you can take an average idea and execute it beautifully, and then you'll have a success story.

Although, saying that is much easier than the hard work you need to put in to make sure your ideas are the ones that are shaping your product's future. Of course, you don't have to generate all the ideas by yourself, as you can also form a team of people to generate more ideas.

For example, your design and development teams need to do just that continuously. Because they're solving problems every day, that means they're creating ideas every day too. You should join them in the idea generation process because no matter how smart you are, you can't think about everything on your own.

5- Conclusion

Idea theft may seem like a problem, but it's not. At least, not anymore. There are ideas everywhere, and the important thing is how you nurture these budding ideas in your unique conditions.

Consult a lawyer about having an NDA in place before hiring staff.

And last but not least, don't worry about people stealing your ideas, it's not going to be a big problem for you. The bigger issue may be that nobody understands it. Even worse, perhaps nobody cares. These are more significant problems, and you should focus on them too.

Your developers may be great at what they do, but it doesn't mean that they can think like you and understand your unique conditions. These are different things.

However, we need a way to spot who's a great developer and who's not. Let's do that now.

What is a Senior Developer?

There are some levels which every developer knows: Junior, Mid-level (Medior), and Senior. This is a way to categorize what a developer can do for you and your project, and we need to understand those levels and why they are essential.

We'll also look at how the terminology can be manipulated and how to avoid fake level indicators. Let's begin.

First of all, it's about the experience. When you start out in development, it's as a beginner, not a junior developer. Being a junior developer may take time. After overcoming the initial challenges, you can work as a junior developer.

Then, the mid-level conundrum rears its head. I haven't seen anyone call themselves a mid-level developer; that's because it doesn't say anything except that you're trying to be a senior developer.

And then there are the senior-level developers. The experienced ones, the ones who spent years perfecting their craft, right? Well, not always. Experience is not only about a matter of time; it's also about overcoming high-quality challenges.

A high-quality challenge means a challenge that is hard to solve. You know it's a high-quality challenge when you feel the fear of not overcoming it. By solving those high-quality challenges, you can build experience.

* * *

I knew junior developers who had been writing code for something like twenty years. Of course, no one could call them juniors, but that's what they were. And I saw senior developers with just three years of experience. As you see, it's almost independent of time.

We need a way to determine the real level of the developer we want to work with. Here are some tips on spotting a developer's level.

1- Look at the Terminology

"Those who understand, teach." It's a famous saying by Aristotle. We can use ancient wisdom as our guide here. If someone is spouting nonsense terminologies to you it probably means they don't understand the concepts they're talking about.

In computer science, there can always be a simple explanation for almost every concept. However, it doesn't mean that the concepts are simple; it's about knowing and understanding the concepts so you can describe them in simple terms.

So, if you encounter someone trying to show off their knowledge by using terminology that's nonsense to you, you can step away safely, and perhaps swiftly.

Great developers don't do that. I know it because I felt it. The feeling of understanding a complicated concept is so good that you'd want everyone around you to understand that feeling. As you know, not everyone around you wants to spend time understanding such a complicated concept until you have simplified it for them.

That's the crucial part of the experience, and that's why it's almost independent of time. Because there are people who can manage to work with complicated concepts without understanding the underlying structure, they just do what they're told, and that's a bad thing to do in a startup.

Because startups need to form things from the ground up, startups need to solve high-quality challenges to succeed, and someone with an ability to understand the challenges to solve them is better for a startup than someone who lacks the nous.

2- Look at the Portfolio

If you want to work with a senior developer, look at their development portfolio. It helps if they have contributed to some open-source projects, but it's not essential. You can take a look at their portfolio and ask questions about which parts were their responsibility.

The things you want to find here are similarities. If you want to work with a mobile app developer and look at a development portfolio, you should look at the apps that have features that align with your app.

You may not spot all the features from some screenshots, or you may not have time to check every app they built, so it would be best if you asked them questions about those features. You are looking to understand whether what they do can help you with your project or not.

3- Look at the Code

Even if you don't know any programming languages, you should look at their code. If it's open-source, it's relatively easy, but if it's not you may need to ask them for a portion of their code.

Just take a look at a portion of their code. Make sure what you're looking at is their own work, because it could just be boilerplate code, and try to understand what that code block is doing.

If you can understand it, it's well-written. If it's short enough to give you a taste of what they do in the first ten seconds or so, it's a good block of code. Here's why:

Computers understand instructions in ones and zeros. They don't understand the code we write unless it's compiled to machine code (ones and zeros) or a compiled program interprets it. So, we're not writing the code we write for the machines. The compilation and interpretation part does that.

We write code for people, not for machines. It's essential—the "people" here are maybe ourselves in three months, or completely different people like you. Extendibility and maintainability are so crucial that we have to write code for people so they can maintain and extend it easily. Modern machines can handle the rest.

Try to understand a random block of code they have written.

4- Ask Why

While you're trying to understand a code block, ask why they did what they did. Give them a chance to explain themselves on their terms. Because you'll eventually need to hire more developers, and the first ones will have to pass on the knowledge about the code they have written.

Just ask the question. No matter what you get as an answer, try hard to understand if it means you are good to go.

In conclusion, it's good for them to have a development portfolio to show you, and if what they do aligns with what you want. Another plus point will be if you got a good and simple answer when you asked them to explain their reasons for doing a piece of work in a particular manner.

Those are some (not all) of a senior developer's qualities. You want to work with one at first because it will make what you'll build together easily maintainable and extendable.

Next, let's take a look at where and how you can find great developers. See you in the next part.

More Quality For Less Money

Finding great developers is a challenging task, but finding great developers when you're on tight budget is harder. So, we need to find a way to work with great people when we haven't got a lot of money.

There are developers all over the world. In an ideal world, you would be able to work with anyone no matter where they are. There may be places where living costs and developer salaries are lower than where you live. If everything goes to plan, you could find someone to work with from another country, and get more quality for less money, right?

Well, not exactly, Even though, in theory, it's a smart move. However, there are some caveats. Let me give you some tips in managing the process:

1- Have Clear Agreements

Ideally, you will have clear agreements. That's because if you don't have them, your processes can easily be manipulated in a foreign country, especially in those countries where life is cheap, and it's hard to find someone with solid knowledge and clear communication skills. If you run into trouble, your agreements can come to the rescue.

No matter how well they speak your language, they will probably have a different cultural outlook to you. That's important, because in some cultures delivering work on time, for example, might not be a priority.

If you want a to build a reliable overseas team, you should write down what you want and get everyone to agree by signing it.

2- Don't Pay Upfront (at First)

Of course, your agreements should have two sides—one for them and one for you. After you state that you'll pay a certain amount of money to someone, it's essential to do it after the work is done, not beforehand.

It's not because everyone abroad is a criminal, it's because your agreements are prepared in accordance with your local laws, and you're responsible and accountable to that local legal system.

* * *

After the first development and payment cycle has been completed, you can adjust it according to your needs. Every method is usable here; just be aware of the first payment.

3- Have a Control Mechanism

Having a reliable control mechanism for your team is a good idea. By control mechanism, I mean a way to see what they do daily, even hourly. It's essential because they're far away, and, without a physical presence, you don't have a concrete way in which to control their actions.

You should at least know what they're working on and whether that is the thing you want or not. Then you can adjust the process according to your project's needs.

4- Know the Local Laws

Knowing the local laws is essential because the people you work with will probably be governed by them if there's a dispute. There may be countries without a well-established law system, which could cause you headaches.

So, know the local law system is an efficient one that could present options in the event of a dispute.

5- Know Who You Work With

This is the most critical point here. If you don't know who you're working with, you're probably working with the wrong person, at least most of the time. That's because you need to know who to entrust with your project's development. As we discussed earlier, the more information you have about who you work with, the better.

In the next part, we'll look at a different approach; the one where I can offer practical help.

* * *

Working with Consultants to Hire Developers

Working with consultants like me may grease the wheels of the process for you, especially if you want to work with developers abroad. That's mainly because they are extremely familiar with conditions like yours and know how to act according to your unique circumstances.

The first benefit you'll see is that you can find higher quality developers abroad. For example, I know many high-quality developers in Turkey, Singapore, Ukraine, India, and so on. By "knowing," I mean I have worked with many of them, and they have frequently delivered quality work for my clients.

That's an excellent way to ensure you have reliable developers. It'll also be cheaper, because, for every country, there is a cheaper one when it comes to living costs. As a result, the salaries are lower too.

It's about how you can spend less while providing the same standard of living for your team. You will also have a better control mechanism because you'll know who you'll be working with and so determine your conditions beforehand. The rest will follow, with your conditions in place.

Let me tell you a couple of real-life stories from my own experience.

An entrepreneur from the Netherlands found me on LinkedIn and asked me to work with him on an app project. He said he wanted it to be developed in his home country, but doing it there was too expensive.

We started to analyze his work and when we have a solid structure, we started to seek out developers. I found two people to work with him from Turkey, who are excellent developers, one for iOS and one for Android. We used Firebase (from Google) as a backend, so we didn't work with a backend developer.

They delivered the work on time, and they're still working from time to time on the app's updates. Because the ones I found were reliable developers to work with and governed by a similar legal system, it was an excellent experience for the whole team and a remarkable journey for me.

Another fantastic journey was with an entrepreneur from the United States, who also found me on LinkedIn. He was already working with a team there, but he thought things could be done better.

Again, we started by analyzing what he had. We spotted some flaws and gaps in the analysis and began to think about how we could fix them. I found a developer and a dev-ops engineer from Turkey and Singapore, respectively, and they worked closely with him for almost two months.

After two months, he offered the developers jobs in the United States. One of them accepted, and they're still working together. They continue to work remotely with the other one too.

Everyone has different conditions, and every condition is unique in itself. You need to analyze your conditions to identify your strengths and weaknesses. Then, with a little help, you can create whatever you want to create and make the world a better place. (Because there's nothing wrong with having ambitious goals, right?)

Let's have a recap of the last two chapters, then; the development phase.

Chapter Recap

We'll recap the previous two chapters together, because they're closely related to each other. Let's take a look at each part.

1- Idea Analysis

We started with idea analysis. Idea analysis is the process for understanding what you have in hand, in broad terms. We spotted the details, created personas for our project, and utilized user stories in order to have an understanding of the users from start to finish.

* * *

User stories can also be used in the development phase, so making them as detailed as possible is a good thing to do.

2- Business Analysis

Then we had the business analysis part. We defined the workload and broke it up into pieces so we could struggle with them one at a time. The main point in business analysis is creating a clear vision for the future of the business we're creating and the market we're in.

Also, it's good practice to have business analysis in place to prepare our ideas for the software analysis process.

3- Software Analysis

Then we moved on to software analysis. You don't need to do this part alone, you can get some help from a professional and they can make this process fairly easy for you. Because it's here that we started to talk in the developers' language.

We included business processes with user & application flows, in a highly technical manner. Again, the more detailed this document, the better.

4- How to Grade Developers

After completing the analysis process, we discussed how to grade developers. Grading developers is important, because we need to have a framework in hand to create a team of champions.

Being objective in this process is important, because when we are looking for a developer we want to find these qualities: productivity, loyalty, an obsession with quality, and a great aptitude for learning.

5- Broad Comparison of the Programming Languages

After grading the developers, we compared some of the popular languages which are out there. Our approach gave us two categories:

compiled and interpreted languages. Each has their unique advantages and use cases.

A quick note here: The language list can and will change over time because software development is a moving field, and we're constantly creating new languages in the process.

6- How to Select the Right Programming Language

Then we talked about four different scenarios, where we built a photo sorter app, a bookmarks website, a bitcoin ticker app, and a hyper-casual game.

We talked about basic projects only, because it gets more and more complicated as they get bigger. There will always be unique features in every project which need to be factored into our calculations. That's why I gave you just a glimpse of what you should do as a starting point.

7- Languages & Tools for Web Development

We talked about popular languages and frameworks within those languages in this part. We took a look at Python with Django and Flask, Ruby with Ruby on Rails and Sinatra, PHP with Laravel and Lumen, and JavaScript in backend and frontend with Express, Vue, and React.

You can develop a backend system for almost any existing general purpose programming language, but in order to be as effective as possible it's better to go with the most popular ones.

8- Languages & Tools For Mobile Development

However, that's not the case for mobile development.

If you want to develop for iOS natively, you're limited to Objective-C and Swift. It also supports C++, but I don't recommend it for iOS development.

* * *

If you want to develop for Android natively; you're limited to Java and Kotlin. It also supports C++, but again, I don't recommend it because it's too low level for a mobile app.

If you want to build cross-platform apps, however, you can use different languages like Dart (with Flutter), and JavaScript (with React Native and more).

9- What Is Full-Stack Development?

In some cases, we don't need to hire more than one developer to develop a mobile app or a website. Full-stack developers are there to help us.

Having them in a team is a bit more difficult, but they can handle both frontend and backend tasks by themselves. The process comes with some caveats, of course, such as, you may need more time to complete the app, because backend and frontend will not be developing in parallel.

10- How to Negotiate With Developers

We began the negotiation chapter with some tips on negotiating with developers. Those are: Know what you have, know what you need, know what they have, and know what they can offer.

It's a no-brainer for negotiations, I know, but we have an advantage right from the start: We have our analysis and their portfolios at hand.

11- Time-based Approach

Then we looked at the different ways of working with developers. One of them is time-based freelance work. I don't recommend it for a project right from the start, but you can utilize this approach from time to time in order to solve small problems quickly.

* * *

12- Project-based Approach

The project-based approach is the go-to one if you have enough in the budget. Have an agreement in place for the development and maintenance phases, and you're good to go.

The main point here is that you need to write down as many features as you can in the analysis. That's because you'll agree on a project scope with the developer, and you'll want to be sure to include everything in it.

13- Sharing Equity & Vesting

Sharing equity is an important part of having a startup, because it causes people to treat it as their own. We discussed why you should have something like ten percent of the company as an employee option pool, what equity you should give to developers and how you should do it.

14- How to Prevent Idea Theft

You don't need to worry too much about your idea being stolen. Of course, you should try to protect your idea but as it enters the public domain it will become impossible.

It's the execution of an idea that matters, not the idea itself.

15- What Is a Senior Developer?

In this part, we looked at senior developers and how they can help you in your journey. We also discussed how you can spot non-senior developers just by talking with them.

16- More Quality For Less Money

We discussed hiring developers from abroad in this section. It may appear tricky, but if you have the right resources, it gets easier and cheaper, while retaining the same quality.

* * *

17- Working With Consultants to Hire Developers

Finally, we discussed utilizing consultants in the recruitment process. As a consultant myself, I gave you some real-life case studies to think about.

Now we're ready for the development phase.

CHAPTER FOUR

Development Process

Time Estimates

It's essential to know when a project will end. It's essential because everything else may be dependent on it finishing. You may have some release plans and you may need to have a definite date to plan accordingly. Or you may have meetings with potential customers which depend on your project going live.

The primary way to establish a specific end date is to make time estimates about each task. That's an excellent way to calculate how long is needed to complete any part of the project. You may decide to release early with the absence of some features and time estimates allow you to plan those kinds of things too.

Making correct time estimates is a challenging task. If you haven't done similar work before, you probably can't estimate the exact time needed. Even if you have experience of that particular task, there'll always be something else to change the timeframe, at least slightly.

So, we have to have some way of making the correct assumptions about the time and tasks at hand. Here are some tips about making the right estimates, which you can use as a double-check mechanism for the estimates made by your fellow developers:

* * *

1- Make Everything Clear

Making everything clear is essential for almost any task in software development. The importance of it comes from telling machines to do something you want. Machines don't understand you as people do. So, you need to give every step of the instructions to the machine to make it work the way you want.

Making everything clear also means you know what you want from the machine. That's also why the analysis elements are essential. The main thing to consider is knowing what you want to do inside out means that giving the machines the correct instructions is easy.

If you know what you'll do each step of the way, then you'll have the ability to see the flaws and gaps on paper (or on the computer). After you see the big picture and smaller details in one place, you can adjust the plan according to your needs.

It's easier to have time estimates for well-defined tasks, because when developers see a well-defined task, they see what instructions they should give to the machines. If you make it easier, they'll give you more accurate estimates.

Providing accurate estimates for each task is critical because there may be thousands of tasks, and a slight deviation in them would cause a big problem afterwards. As you wouldn't want that to happen, make everything crystal-clear in the analysis phase.

2- Sort Everything in the Correct Order

There may be occasions when the time estimates are correct, but they're not usable, and there are a number of reasons. One of them is the incorrect sorting of the tasks. If you don't sort the tasks in the correct order, they may not be ready for development.

For example, if you have a task that entails changing a profile image in a profile page and you put that before the login and signup tasks, it can't be estimated in the correct context. That causes a slight difference in the time estimates because having tasks in context makes it easier to think about them in conjunction with the surrounding tasks.

* * *

This also ensures that everyone is on the same page. If you have one task from here and one task from there, you can't even talk about what's next and how long it would take.

So, sorting everything into the correct order is the way to go. The correct order means an order which aligns with the user flow. Users are the central part of your apps, so play along with them whenever you can.

3- Know Who's Involved

Another way to end up with incorrect estimates is not knowing who or what's involved in the tasks that you're estimating.

For example, if you're developing the frontend of the application and you need to make it talk to the backend from an endpoint, you want to make sure it's there before you do anything. If it can't be done, you may at least want to know what the exact structure on that endpoint is.

It's almost impossible to know everything in the project if you're not doing it all on your own, so you'll need to figure out a way to align the time estimates from different parts of the project.

If you have frontend estimates in a web project, you're halfway there. You need backend estimates, too, if you don't utilize an app backend platform. And you also need to align those estimates with each other because some tasks may be dependent on those in other people's hands.

So, you need to clearly define who's involved in the task at hand and who's responsible for which parts of the project.

4- Have Buffer Times

Last but not least, you need to have buffer times to allow for things going wrong. Your computer may be broken, the internet could be running slow, or your compilers might decide to stop working for an hour without any reason.

For covering those kinds of problems, you should have buffer times. They make you ensure you're aligned with other parts of the project no matter what.

For example, you can safely give an estimate of two hours for one-and-a-half-hour's work. It makes calculating times more manageable, and it ensures you can do the task well and test what you did.

Having buffer times is a great way to stick with your time estimates and not redo them repeatedly. It takes time to redo the estimates if you miss some of them along the way, so having buffer times is always faster.

Just don't exaggerate them. All you need is a little bit of extra time to make sure everything is working correctly.

Now let's discuss how we can talk the same language as our fellow developers. It's essential because it makes working with each other more pleasant if you understand each other, but if you don't then the relationship can become more challenging over time.

Talking the Same Language with Developers

For non-developers, development may seem like black magic. The developers do some things on the computer, and everything moves along perfectly. Well, that's not what the process looks like.

All the things we call "technology" have trial and error behind them. That's how some developers wrote the code for almost everything we see around us in the modern world.

Development is a challenging process, but it's not black magic; it's all code written line by line. Anything seems possible in theory but not in practice. In practice, there are some constraints. Constraints like time, budget, and knowledge. Constraints that everyone has.

Developers are no different. There are countless things to learn if one is to become an expert in any development field, so they have to work hard to keep it up. Sometimes, they need to work like the machines they write the code for, enabling them to learn how to do some seemingly simple thing.

So, talking the same language as developers can sometimes be challenging. Let me give you some tips on how to communicate effectively with developers. By the way, they're as human as you are, so the primary communication skills you have will be enough for day-to-day activities; I'm talking about technical matters here.

There will be problems in the development process, no matter what, and your task as an entrepreneur is to minimize the impact of communication problems. That's why it's best to talk the same language as developers. Here are the tips:

1- "How" Can Be as Important as "What" or "Why"
What to build is not a problem for experienced developers, but deciding how to build it is. Experienced developers also know that they're not capable of doing everything at once, so they have to plan to achieve their goals.

Developers already know "why" (if you've done your job well). Your job is to give them at least one "why" to work on your project. They also know what to build because you created documentation beforehand, and that's why you're a great tech entrepreneur.

After giving that much to the developers, you can expect them to figure out how to build the thing you want, right? Well, not exactly. You'll need to discuss the "how" part with them too, and because what they'll do is highly technical, you need some prior knowledge of what they're talking about.

Let's see how you can easily acquire that knowledge by talking and listening to them.

2- Listen to Them

That's the second tip. It'd be best to listen to your fellow developers carefully before sharing your opinions on the development phase. It's important, not because you lack technical smarts, but because the developers' job is to think at all systems levels.

Thinking about systems is not an easy task, nor is explaining your thinking on the subject. It's like thinking about one thing, but without thinking about it directly.

So, it reflects on the communication between you and your fellow developers. They can help you understand a concept you're not aware of, but only by showing you something else.

For example, you can describe "Object-Oriented Programming (OOP)" by comparing it to a car. Just as OOP has properties, your car has them too, like "maximum speed." And just like OOP has methods, your car has methods too, like "accelerate."

Using metaphors is a way to explain a somewhat complicated concept, but you need to listen to it from start to finish to understand the concept. It's the flaw in systems' thinking; you can't understand by looking at half of it. You need to listen from start to finish.

3- Do Your Homework

Doing your homework means two things:

First of all, you need to work on other parts of the business, ensuring everything is operating well. That's because the company's overall motivation affects the communication between entrepreneurs and their teams.

If you continue to work on the other areas of the projects, and not abandon them, it becomes easier to communicate with your fellow developers.

Second, know something about what they're trying to achieve. It may change over time, and it'd be best if you have some knowledge for every situation they're in.

* * *

You're the ship's captain; you should have a route at your fingertips to guide them continuously. However, there will be times when you may need to acquire some extra knowledge.

You can acquire knowledge just by talking to them, and it's the best way to make sure what they do aligns with your vision. It's hard to keep up with the vision because day-to-day tasks will take up the time needed to think about the bigger picture.

In those times, you'll be responsible for continuing to think about the big picture.

4- Allow More Time if Necessary

Time is an exciting topic to discuss because, as everyone knows, while we all have time, none of us have enough of it. This is the point where you should relax and make some time for yourself.

Just because things get delayed, your team may not always be responsible. You might be causing delays too. So, give yourself enough time to reflect on what you do and what you will do next. It's essential for you to be all in, but you can't be in that mode continuously. At some point, you'll need breaks.

Your team will need breaks too. They may have been working hard on the project for quite a while, and that might be making it hard for them to think clearly about what they're doing. So, forget about the time estimates for a while and give them time to kick back and decompress.

Some problems can't be solved by working on them exclusively. Instead, a solution might be found by you doing a completely different thing and not even thinking about the problem. Developers are no different.

So, the next time you see one of your fellow developers seemingly doing nothing, think again; it may turn out to be a good thing for the project.

* * *

Let's discuss how we can manage the project with the developers.

Managing the Project Together

We'll discuss project management techniques in upcoming chapters, however, I want to spend some time on it now. That's because if you can manage a software project with a developer or a team of developers, it offers up a number of opportunities.

For example, if you can experience the whole process from start to finish with your fellow developer, it greatly increases your understanding of what's going on in your app. It might seem tricky, but I'll give you some tips on how to benefit without spending too much time on it.

First of all, managing a project is an arduous task. You need to manage every aspect of it, and it can grow into a much more complicated beast than it was at the beginning. The more you delegate the actual management function, the more time and energy will be freed up, enabling you to market your business to the outside world.

That's why you should consider managing the software project "with" the developers, not watching over them as a supremo while they work on the project. The difference between the two is the delegation of some of the management tasks to the developers.

Managing software development processes is easier if you divide the tasks into little pieces, and if you're working with an experienced developer that's something they will be aware of too. You can allow them some leeway in the management of their project, which also gives them the motivation to finish the app - and motivation is always needed.

You can have an opinion on everything, and these opinions may be necessary for the work to be done, but developers can have opinions too, and it's better to consider them in the project management phase. Let me give you some tips on managing the project with the developers.

* * *

1- Make Sure You're on the Same Page

One of the tips I can give you is to make sure you align with your team, at least for most of the time. It's essential because it allows you to iterate on your plans, polish them, and keep track of your progress.

The primary way to make sure you're on the same page with your team is by regularly talking to them. If you talk to them daily, attend short, daily meetings called "scrums" (we'll explore that later), and listen to what they have to say, you can make sure you're aware of what they're doing and how it can affect the overall business.

If you understand what they're doing on an incremental basis, you can also adjust deadlines and solve problems they might face before they even arise. It's one of the essential qualities of an entrepreneur: Seeing there's a problem on the way and solving it upfront. It's a muscle, and you can develop it.

All successful entrepreneurs are problem solvers. They solve not only big problems but also the smaller ones. Being on the same page with your team helps you solve more daily problems upfront.

2- Transparency is Important

We already discussed why your team's transparency is essential. However, what I want to emphasize here is your transparency, which is also vital. Being transparent is important because it's essential to the building of a loyal team.

Being transparent can be hard sometimes, especially when there's no one to remind you on a regular basis. Because you're working for yourself, there's no one to force you to be transparent and show you what you should be doing with your team.

However, the good news is that if no one is forcing you then the onus is on you to force yourself. It's a good thing because it's a way of earning your team's loyalty and your progress can be a motivator.

* * *

Don't reserve your transparency for your successes, be transparent about your failures too. Don't think doing so will demotivate your team, because most of the time it has the opposite effect. Sharing both your high points and your lows will make them feel important, and they'll be there to help you in hard times.

3- Decide Who's Responsible for What

Setting the areas of responsibility is essential, especially if you're a growing company. If you don't set the responsibilities clearly enough the chances are that there may be problems surrounding who's responsible for releasing the app.

If everyone is waiting for each other to do something, it's known in development as "deadlock." There can be no progress in deadlock, which is a type of failure. You can't create movement within a deadlock situation without changing the underlying causes.

You should at least be deciding who's responsible for the critical actions. For example, if you have three developers you need to give one of them the responsibility to release the app they're developing.

If your team is waiting for you to arrange the privacy policy, you should be keenly aware that you're responsible for the current problem and there can be no progress before you solve it.

If everyone knows their responsibilities, you can make progress more manageable and more efficient. And making progress is one of the most important things for a startup.

4- Give and Take Feedback

Finally, you should give your honest feedback to your team and take feedback from them. It's like being transparent, and it's crucial for your team to be a real team, not just a group of people.

Giving honest feedback can be problematic because it forces you to think hard about the situation. However, it's essential, because if you don't say something in good time it can be wasted or simply irrelevant.

* * *

Delivering honest feedback is important in generating a healthy working atmosphere and a happy team. If you don't share feedback, it can be more challenging for you to keep up with what your team is doing.

The opposite side is also essential. You should encourage your team to give you honest feedback, even if it's not good. That's important because it makes them develop a sense of ownership of the project.

Now it's time to explore some technical aspects of development, starting with storing the codes your team writes.

Where to Store the Codes

You may have thought about this before; you need somewhere to store the codes which your team has written. Here are three reasons why it's essential: First, everyone needs to be on the same page when developing something. Second, for collaboration between developers, and even with their different computers. Third, you need to have the ability to roll back when you make a mistake.

For those reasons, and there are many more, developers created version control systems. A version control system is a program that gives you the ability to save each action you do as a "commit," with the option to include a message. It allows you to track what you and your team are doing and when it was done.

Also, version control systems allow a team of developers to work on the same code base, and without them working on the same code base would be a nightmare. For example, if you want to add a new feature, and someone else in the same team wants to solve a bug on the same code base they need to make sure they're not breaking the whole project. Version control systems are here to help us insure against such issues.

There are many different version control systems available and in widespread use, however, we'll concentrate on just one of them: Git. Git is the most commonly used version control system, and every developer you work with should know how to use it. So, it's safe to say it's the thing you'd want to use.

Using Git is reasonably straightforward. You initialize a "repository" on a folder in your system and start tracking the files in it. If you add files to the repository to track them, it tracks them. If you change something in a file you're tracking, you can commit the changes you're making.

You'll then need a way to share the code you write with your team or store it somewhere on the cloud. Here are the Git services: You may have heard of Github, it's a service you can use to store your Git repositories and it has many benefits, including code search, access control, pull requests, and many more.

Using a cloud repository management service is a great way to store your codes. You can control who has access to what, and you have the luxury of seeing what has been done on the repository so far from an excellent user interface.

There are also other cloud repository management services, such as Bitbucket and Gitlab, and they have some advantages and disadvantages when compared to Github. You can check them out and see for yourself. It's very much a matter of personal preference since they can all do the same things, but in slightly different ways and with slightly different user interfaces.

You should use a version control system, preferably Git, and you should store your code in the cloud for easy access. If you want your code to be on your servers and not on a cloud service, that's also possible.

Git is manageable through your servers, and you can have Bitbucket or Gitlab on your servers too. So, you don't need to have a cloud service to store the codes, you can use your servers. However, I'm suggesting that a cloud service is what you need to store your codes in most cases.

Let's look at the usage rights of the codes.

* * *

Usage Rights of the Codes

This is a highly debatable topic, and I'll do my best to walk you through what's on the table when it comes to the codes' usage rights.

First of all, codes written by anyone, especially for the company, are part of the company's intellectual property and are owned by the company and nobody else. However, that's not where the debate starts.

The debate starts with the question: "When do the codes become intellectual property?"

There are two main approaches to answering this question. In the first one, the code becomes intellectual property as soon as you pay for it. In the second, the code is intellectual property as soon as someone writes it.

There may not be that much of a difference at first glance, but you may encounter situations where there's a noticeable difference, especially in developing countries. In established economies, people tend to pay on time. However, in developing countries, it may be different. As a countermeasure, developers and creatives living in those countries had to find a way to protect themselves from theft.

For example, you start to work with a team of developers abroad, and you want to see the code in the middle of the first month. If you don't set your agreements correctly, they can refuse to give you access to the repositories where they store the codes.

Because they may not trust you to pay them on time or something like that, their actions might be justified. It gets complicated when you want to have the codes in place as your company's intellectual property.

To solve this complicated problem, we should have a clear strategy, and I suggest three things:

First, have a strict rule that ensures nobody can store the codes outside your cloud repositories. You should have sole control of the repositories on the cloud. No one should set up their repositories with the codes written for your company. You would see the codes immediately after someone writes and commits them.

Second, have clear access controls to the cloud repositories. Learn how to manage the branches and maintain the ownership of at least the "master" branch. The master branch is the main branch of a Git repository. You can restrict access to the master branch, ensuring no one can enter your master branch except you. It's an excellent way to store the codes if you know what you're doing, so spending some time learning more about how Git works would be worthwhile.

Third, pay everyone full, on time. It's so important, that I don't think it needs any further explanation. If everyone does the best they can the world will be a better place. Please do the right thing and don't disappoint those you're working with.

Now that we have a way to store the codes and maintain the ownership of them, let's take a look at project management techniques. There are many of these in existence, but we're going to look at just a couple.

Project Management Techniques Overview

When it comes to project management, there are many techniques out there. Because it's hard to manage a software project from start to finish, some smart people have developed methodologies for doing just that.

Let's take a look at some of them.

Waterfall

The Waterfall project management technique is the most straightforward way to manage a project from start to finish.

It starts by writing down everything you need to have, in detail and in order. You work on the project linearly for a set end goal. Everything you need to do to release the project is already set, and you just need to do what's needed.

With Waterfall, you don't change the project in the middle of the process. You wait for the project to end and then iterate it or write from the ground up if things change.

It was traditionally used in the software development industry before Agile methodology was born. You set a goal, arrange everything towards that goal, and don't change the goal before completing it.

It seems a great way to stick to your values, right? Well, not always. In the real world, requirements can change. For example, you suddenly realize you need a feature, and you need it now. So, you have to change the requirements along the way.

However, this is not good for the Waterfall technique. Because everything is incrementally designed, you may need to redo the plans all over again just to accommodate that feature, and that could cause you headaches.

You can use Waterfall methodology if you have a small project and you're so familiar with what you're doing that you won't need to change anything during the process. As another option, here comes Agile.

Agile

Agile methodology was created by a group of highly talented people back in 2001. It's an incremental and iterative approach to software development.

It came into being because Waterfall was slowing down the software industry, and the world needed a way to handle things when requirements changed.

Agile methodology allows you to manage your project in short cycles called sprints. In each sprint, you set the requirements, plan, design, develop, implement and review. Sprints usually take a week or two.

This allows you the luxury of iteratively testing your product's features, changing them to meet your customer's expectations, and releasing.

Of course, you can't iteratively release an app on the app store at first, but you can release a testable version to a small group of testers to find out what you did right and what you did wrong.

Comparison

Let's compare some of the aspects of Agile and Waterfall methodology:

Agile is incremental and iterative but Waterfall is not. This means, with Agile, you can fix something in the middle of the project. Waterfall doesn't allow changes to the requirements.

In Agile, you test the product continuously but with Waterfall, you test after you develop. You can't see the errors without testing the product, so testing earlier is better.

Agile is broken into pieces to make the whole process easier to manage and develop. Waterfall gives you one whole piece to work on.

In Agile, you and your customers have the opportunity to change the requirements and add or remove features along the way. You shouldn't do that with Waterfall unless you want to start from the beginning.

Waterfall is strictly structured, everything is in place from start to finish, and everyone knows what they'll do after they've completed each task. Agile is not that structured; it depends on iterations and getting something out there to test and refine. Not having a strict structure is a good thing because you can adapt to changes along the way.

You can know almost precisely with Waterfall when the

development will end , no matter how big the project is. In Agile, you can't always know precisely when the development phase will finish because you'll be iterating along the way. Iterating makes it harder to estimate the project's duration before starting it. Although you can do it easily with smaller projects, it's not that easy on bigger ones.

You need to have clear, written documentation that includes everything the software team needs when using Waterfall. In Agile, you don't need documentation from the start, you can develop it as the project proceeds.

In Agile, the project designers and testers work closely with the development team. Everything is designed, developed, and tested in short cycles, so everyone gives each other the necessary feedback before it's too late. In Waterfall, the designers, testers, and developers do not work together, but in different phases of the project with their teams. It makes it even harder to implement change.

In Agile, you need to approve the result of every sprint. It's a good thing because you make sure the project is going forward and can change anything you don't think fits. In Waterfall, the acceptance process starts after the development and test phases. So, you'd need to wait before you can see a working product to test and accept.

In Waterfall, you need to plan everything upfront, which can be considered a strength until you need to change something. Then you may need to plan all over again. In Agile, the minimum of upfront planning is required. You can plan the next sprint, or couple of sprints, and start developing them.

In Agile, designers and developers can work closely in the analysis and planning phases but with Waterfall they're separated from analysis and planning, so they'll only have a final plan to execute. It's a rigid state for a startup to be in.

Consequently, you can use Waterfall in small projects where everything is already set, and everyone knows what to do. Otherwise, use Agile. Agile methodology is one of the best things that happened to the software development industry. Learn it well, and use it.

* * *

In the next part, we'll take a look at the "scrum," the heart of Agile project management, and we'll discuss how we use Agile in our projects.

Agile Project Management & Scrum Methodology

Software development and project management methodologies are there to help you build your project from start to finish. You can start a project and finish it without using any of the available methodologies , or create your own methodology, but, it's nothing more than wasting precious time for an entrepreneur.

If you want to have a stable work environment, it'd be best to use some project management methodology. Agile is the one I use, and I suggest you use it too. I'll explain how to utilize Agile methodology in a little detail:

What is Agile?

Agile is a way to manage your project with short cycles. You describe what you need, create it, test it, iterate it, accept it, and then move on to the next one. Of course, the cycles are connected because they're in the same project, however, they're different blocks of work, and you should work on each of them separately.

Sprint

Those cycles are called sprints. Each sprint has its planning meeting, working phase, daily meetings, review meetings, and retrospective meetings. Let me explain each of them.

The planning meeting is where you get your team ready for the upcoming sprint. You have a thing called product backlog, in which you store all the tasks and user stories from the software analysis we discussed earlier. In the product backlog, you can see the things you need to do next, and with the help of your team, you decide what to build in the next sprint.

* * *

After planning the sprint and ensuring all the tasks are in order, you move your tasks to the sprint backlog. It's different from the product backlog because it's there for you and your team to select individual tasks. You don't select individual tasks from a product backlog, the sprint backlog is made to select and work on tasks one by one inside a sprint.

Then you start the sprint. The typical duration for a sprint is a week or two. If you need more time, divide the work; if you need less time, merge the work. After you have started the sprint, everyone knows what to work on for at least a week, and they've selected which tasks they will do to make the sprint a success.

Scrum

Here comes scrum. Scrum is the core methodology to utilize sprints. It's also a meeting method where people answer three questions: 1- What did I work on yesterday? 2- What will I work on today? 3- Are there any obstacles I need to tackle?

You can give everyone a couple of minutes to answer those questions one by one, and that's all you can do in the scrum. That's why we do it standing and not sitting, because if you have a standing meeting, you end it early. You should end the scrum meeting after ensuring everyone is on the same page.

There are three critical roles in a scrum meeting: Scrum master, product owner, and scrum team. However, in small teams, the scrum master and product owner roles may merge. You will probably be both of them.

The scrum master is responsible for the scrum meeting going well, making sure there are no obstacles, and everyone's there simultaneously - preferably in the same place every day. The product owner is responsible for the product backlog, grouping the tasks, and bringing them to the sprint. The scrum team is the development team that does the actual coding, designing, and testing.

* * *

Kanban

There's a useful technique for smaller teams called "kanban," which serves as a replacement for sprint cycles. Using kanban is a way to get rid of the structure that scrum and sprints have to offer and give a bit of room for small teams to work on.

Kanban is selecting a group of tasks from the product backlog, putting them into a "to-do" list, and moving them one by one to "in-progress," "in-review," and "done" lists. That's the essence of kanban, it's simple and usable for any project. It's better when using small teams, consisting of less than six or seven people.

Comparison

There are differences between scrum and kanban; let's take a look at what they are and how can they affect your workflow:

In scrum, you need to select a group of tasks and you need to fit them into a sprint. There are no particular restrictions about how many tasks you can select for development in kanban. You can select two tasks, or twenty, it's all up to you and the speed of your workflow.

In scrum, you need to complete the work you have inside the sprint before the sprint ends, but there's no need to have a time commitment on tasks in kanban. Having estimates and time commitments is reasonable, but the methodology doesn't restrict you in any way here.

In scrum, you should prioritize your backlog and select the best ways to deliver the most critical tasks first. In kanban, you don't need to have explicit prioritization. You can select whatever tasks you want to be done and start working on them.

Scrum has a number of charts, such as the burndown chart. The burndown chart gives you the birds-eye view of the tasks at hand to make sure you can see the project status. In kanban, you can also utilize a burndown chart with some tweaks, but it's not common or particularly necessary.

In scrum, the board you put your tasks on resets after each sprint because each sprint is a new workplace, and each sprint has its task list. There's just one constant task list in kanban, and you can do whatever you want there.

Which One to Choose?

In my personal projects, I use kanban. You can use kanban in your personal projects, too, when you need task tracking. It gives you the ability to see where you are in the project and gives you the freedom to select what you want to achieve.

In small projects or with small teams, I use kanban too. It's easier to modify and adapt kanban tasks, and small teams don't require large systems to be in place.

In larger projects or with larger teams, I use scrum. It gives a sense of structure and ensures everyone is on the same page every day.

In the next chapter, we'll look at sustainability in development.

Sustainable Development

Sustainable development is a topic that can easily fill a book all on its own. It means writing code that doesn't need to be re-written repeatedly.

Every type of software needs to be re-written from time to time. That's normal because technology is going forward, and new improvements are coming along every day. Which is not significant from a sustainability point of view, but the important part is whether you can manage to use the code for longer than your competitors.

It's important because there will be staff turnover in your development team. People will come and go. If the first ones to write the code do their job well on sustainability, you'll be good to go. Otherwise, you'll have difficulty finding someone to take the codebase and move along with it.

* * *

Sustainable development is a broad topic, but I have four tips for getting started.

1- Code Quality

If we were writing the code for machines, we'd write it in ones and zeroes. However, it's not the case. We're writing code for humans to read. That's why programming technology is moving forward.

So, we need to write human-understandable code, and we need to do it every time. It's as simple as giving variables and functions exact names, restricting their scopes, and making them understandable in seconds.

The main point here is if any part of the team will be subject to change, nothing should be affected in the codebase. The next developer's reaction will be directly linked to the code quality. You don't want everyone to change after seeing your codebase.

Sustaining code quality is a challenging task. It's hard because it needs multiple things continuously. First, you need a developer who understands the programming concepts, software design patterns, and clean code principles.

You'll need them to be continuously motivated to care about the code quality, because making mistakes is easy if you don't insist on the writing of quality code. For example, you can name a variable "x" and use it in ten different places for different purposes, but no one will understand what "x" is at first glance. Naming it "database-connector" instead of "x" requires motivation. Because they both will work well for machines, it's humans that need to understand the difference.

2- Room for Improvement

One thing as important as code quality is the room for improvement. It's about extending your program's functionality and how easy or hard it is to do.

For example, if you're developing a photo sorter app and want to add a share button to share the photos in multiple places, that shouldn't be tricky. But it can be challenging if the software is not designed to extend.

The main difference is time. It's common to use shortcuts when you feel time pressure. There are many ways to do the exact thing you do with the code, so selecting the right way from the optimal ones can be tough.

It may take time, but if you can do that it'll give you the ability to extend without doing extra work. At the end of the day, it means not spending time but earning it.

So, the best developer isn't necessarily the fastest, and the difference is evident in time estimates.

There is no black magic in development. If it seems unreasonable for a person to code something in the timeframe they're estimating, it may mean two things: First, they may have the codes already written for another project, and they're planning to use them; or second, they don't know what they're talking about.

Think about developers as humans, just like you. If you couldn't complete an app in three days from start to finish, don't trust someone who says they can. You may not know how to code, but you're capable of spotting the difference between reasonable and unreasonable.

3- Separation of Concerns

It's the technical part of the discussion. It's about the things we call "functions" or "methods." There will be some functions in your codebase that do something, and they need to be designed appropriately.

Proper design means doing what it should do and nothing else. If a function wavers from that requirement it is not useful.

Testing it doesn't require any programming knowledge. Just take a function that your fellow developers wrote and have a brief look at it.

If you understand its purpose in five to ten seconds, without any prior knowledge, then it's a great function.

Most of the time, functions can be divided into smaller parts. However, dividing them might be challenging. If you created a hundred-line function, then dividing it means you need a way to make sure it works properly after you divide.

We'll discuss software testing in an upcoming chapter. For now, be aware that it's important to write code that humans can understand without prior knowledge.

4- *Continuous Audits*

Auditing code is demanding for both parts of the equation. You can't give the whole codebase to someone to code, do something else, and never look back again. It's not about trust; it's about being human. Being human means you can make mistakes along the way. It also means that there are times when you will get tired.

There can be many reasons why the quality of the software you write is not sustained, and they increase as the project gets more complicated. Even though the coded quality may not be in danger you need to find a way to audit it from time to time.

It's not about judging the developers on what they did, it's more like solving problems with them. Being at the heart of the software is a challenging task and you can make it easier by having someone check what is going on.

This means we need a way to audit the codebase continuously, so let me introduce you to "code review." It's what we'll cover next.

Code Reviews

Because performing continuous audits is an arduous task, there was a need to find a way for it to happen without causing pain to non-developers. Thankfully, some smart people came along and invented code reviews.

A code review is the process of different developers checking the committed code and approving or disapproving it. It goes like this:

A developer commits some code to the code repository, which you're keeping in the cloud or on your own servers. Then, it immediately enters a state where a review is needed, and an email (preferably) is sent to the second developer to make them aware of the fact.

The second developer then checks the changed parts in the commit and approves them. If everything looks OK, they move on. If something seems wrong, the reviewer asks for a correction and disapproves the commit.

In theory, non-coders can do code review but it's complicated, and tough to do after the project starts. For that reason, I prefer a third party who knows about development to look at the codes and review them daily.

Just because it's difficult doesn't mean you can't do code reviews yourself, but you may need a little help. Here are some tips to help you find your way through the code review.

1- Objective Approach

First of all, it's not ideal for someone to review the code they have written. That's because the whole code review process is about looking at the code with a fresh pair of eyes. If you wrote the code yourself, you are already familiar with it which might make you biased.

It's like writing, editing, and proofreading a book all by yourself. The whole point of editing and proofreading is to spot your mistakes from an impartial viewpoint. That's why, if you write a book, you should use a separate editor and proofreader.

The same goes for the code review process. If you have just one developer in your team, it means you should review the code they have written. You can't be sure their approach is objective if they're checking their own work.

Having an objective code review is an excellent thing for a software product. It reduces the number of bugs and allows you to release with some level of confidence. Of course, you can't make sure everything is 100% in place and correctly ordered, but you know at least two people saw the code and approved it.

2- Comments in Code

The codebase doesn't only consist of codes. There are also comments in code, which makes it easier for someone looking at it to understand what's going on. You can comment on your code however you like, and the comments won't be seen by the compiler or interpreter machines. People will see the comments and that's why they're there.

There are different approaches to commenting on your code. Some developers say you should craft such great code that you don't need comments. Others say the comments should be there in every line, describing what's going on. This debate depends on personal preferences and either side can point to pros and cons.

Let me tell you about my way of using comments in code. I want to see documentation comments before every "function" or "method" (they're the same, it's just the scopes they have.) A documentation comment is about describing what the method is doing, and it can be parsed via documentation engines. So, you write your code with your code documentation; that's a great thing to do for those who come after you.

You can also come after yourself. It's like focusing on a different thing for a month and then returning to what you have written for a fresh look. You'll need a way to find out what the code is all about and comments will you.

3- Selective Reviews

If you don't have enough time, energy, and budget to have extensive code reviews, you can also do selective reviews. A selective review means you review some parts of the code to ensure the developers are on the same page.

While it's not a full code review and it doesn't have all the advantages of one it does have some of them. One of the advantages is that it forces the developers to write the code for someone else to read. It's essential because, as we discussed earlier, we write code for the human eye.

Machines don't care about what we write as long as we give them the right directions, but humans do. Because humans may need to adjust, improve, and change the code. Machines don't care about changing the code; they just do what you tell them. In contrast, humans think. They think about how to improve the product and how to do more with it.

You need to write the code for other people to see and improve and selective reviews can be a good way to achieve that, but they're not extensive reviews, and they won't give you as much confidence.

4- Having Standards

Having standards makes the code review process easy. You can have company-wide coding standards, such as standards in a variable in function naming, maximum file lengths, commenting, and so on.

It's the reviewer's job to make sure everything agrees with the standards you have in place. It makes the reviewer's job a breeze because machines can also help find the errors if you have clear standards.

For example, if you utilize "camelCase" in your variable names, a machine can quickly check whether your naming seems correct or not. So, having standards is a good thing for both machines and humans.

Humans can also grow muscle memory to check the code for errors if there are enough standards. Because if a clear structure is present every time you look, you can easily find your way around it. It also helps the reviewer to perform the review quickly and confidently.

Next up, let's take a look at "quality assurance."

Quality Assurance

Quality assurance (QA) is a popular term in almost every industry today. Software development is no different. We need to find a way to make sure we're creating a great product.

Finding a way to make sure about the quality is like grading developers, it has to have a subjective approach. Measuring the quality of the software is not a completely objective process - let me explain.

For example, you want to have a framework to calculate the software's quality. You can measure the time your developers spend on developing the product or the lines of code they produce, but these quantitative measures won't show you the work's quality aspect.

However, if there's only a line or two it doesn't matter if it's quality code or not. So, quantity matters too. We need to consider qualitative and quantitative measures together to calculate a final result.

We need a process to do that, and we've already discussed some of its factors, like code reviews and time estimates. Let me give you some tips on making sure of the quality of the code your team is writing.

1- Have a Code Review Cycle

When we were young and eager to get to work on some software, we were coding like crazy, so crazy that we weren't aware that we were creating garbage. It's a great way to learn software development, however, it's certainly not the way to ensure your work is of high quality.

If you do something on your own without a feedback loop, you may think that you're doing it right, but you may well not be. It's like writing a book and having some errors in it; it's completely ok, but someone needs to check for mistakes, and you need to solve them all before you publish.

In software it's a bit harder because you already have a feedback loop: the computer. You code and you get the results immediately, most of the time. After getting the desired results, you feel like you're on the right track again and again.

This process, what we call "developer testing," may give you a codebase without obvious errors but there's more than that. You need to find a way to make sure everything in the project aligns with the project and the team.

In these situations, you need code review cycles. It's simple and can be done by a couple of developers in your team. Someone writes, others review, and that's the core process. You should have a rule that states all the production code should be reviewed by at least one developer other than the code's actual author.

If you manage to do that, you'll have a great code review cycle. Some teams live without code reviews, so it's possible, but I don't suggest you try it. It's because we're writing code for humans, not machines, as we discussed previously.

2- Be There When Needed

Your development team will need you as an entrepreneur. Because, while coding, developers shouldn't "think." Of course, they're thinking about the code they're writing, but I mean they shouldn't think about any business decisions.

There will be many business decisions to be made when it comes to the development phase. That's why one-person teams act fast. The main point here is determining who should be there to make those decisions without causing a delay.

For example, a business decision might be about the alerts you show to the users. Chances are, your designers may not design the alerts you'll have in the project, but there are some system alerts and some non-system alerts in almost all apps in real life. It brings up the need to decide how you should utilize the alerts, and someone must make that decision.

Or another example: you may need to decide the default language of your app as it may not always be English. How your app's name should be written in the app is also something that needs to be decided.

These are all micro-decisions that someone needs to make so things can move on. You'll encounter a ton of them along the way, and when they crop up you should be ready to make the final decision and so allow the team to make progress.

They seem small distractions, but, believe me, they slow down the process of development. So, be there when needed to make clear decisions about the process. Even if your decisions don't quite hit the spot at the first time of asking, you can always change them later on.

3- Perform Random Checks

There should be a way for you to look at what's been done in the codebase. However, it's not about you coding or even understanding the code. There should be a way for you to check whether everything is in place and making progress.

There is a way to do that for the product itself, and we'll get back to that in the upcoming chapters, but we're talking about the codebase here.

I have found that the best way is performing random checks. Making those are easy with this four-step process:

First, open the code repository you have on the cloud and then open the branch your team is working on. Second, check the timespans between each commit and try to find consistency. Consistency wins in software development, so it's essential. Third, you get into a commit and try to understand it. If you can understand it, so will the developers you'll work with in the future. And fourth, ask random questions about what's going on in the file you're looking at.

In this way, you make sure your fellow developers know their commits. Because there are times when developers just throw a bunch of code into the repository without proper descriptions. While that's OK for the beginning of a project, it's not acceptable on an established one. That's why we do code reviews, too; we're trying to find a way to deal with increasing complexity.

Now let's move on to the part where you can track what's being done. We're going to look at how you can track your team's work and the time spent on the project.

Work Tracking & Time Tracking

In 2015, a friend of mine, a wellness app owner, called me. He said he was in trouble with his development team abroad and asked me for help. When I arrived at his office, he seemed a bit frustrated.

He had worked with the team for more than six months and they released an app together. Things were perfect until they were not. The work started to slow down, and when he asked the team to explain, they replied, "It's getting complicated." He thought, "That's reasonable," and moved on.

However, things weren't how they used to be. The deliveries started to have obvious bugs and that added to the release time. He wasn't happy, because he had to tell the shareholders what was going on, but the problem was that he didn't know.

When I arrived at his office, I immediately requested the code and took a look at the commits on the project. When looking at them from a time perspective, they started very well. They had plenty of commits in a short period and they released the first version of the app in almost no time. Then, they almost stopped working.

Next, I looked at the code itself. The developers had created a system where everything depended on the servers they were running. They put some black boxes inside, some functions to call their servers to get some critical results for the app to use.

When I asked my friend for the code they had for the servers, he told me that he didn't know. There was a fraud going on. The team, which was in a faraway country in Asia, had created a system where they worked however they wanted, released whenever they wanted, and got paid in full with fixed timeframes.

They were actually blackmailing the entrepreneur with the app they wrote. It's not acceptable, and I had to work my socks off to form a new team. It took me more than a week to put everything in place and get it up and running. I'll tell you, that was hard.

The main point here is that you should trust who you work with. However, it doesn't mean you blindly go your own way and forget about the development process. It's quite the opposite. Everyone in the team should be aware of what's going on, preferably made aware daily in small teams. And if you have a big team, form smaller ones within it to achieve the same result.

Work & time tracking are tools for you to utilize to create a workplace where everyone knows what's going on. Starting with yourself, you should track which task took what amount of time. After that, you should track what the task was and why it was important for the company.

It might seem to be overworking the issue, tracking everything you do, but there are systems to help you track what you do and when you do it. It's a straightforward process when you get used to it.

After you have started to track yourself, start reporting back to your team. Because you're a team and you need to stay together around the same goal. Don't be a boss, be a leader and try the onerous things out on yourself first.

Having done that, you can ask your team about what they are doing, why they are doing it right now, and what they will be doing next. Tracking doesn't mean knowing every detail about the process, you just want to know what's going on every day.

* * *

There are tools for freelancers to use, tools that can take screenshots at random intervals, to be used as proof of work. If you're working with freelancers or an overseas team, you can utilize those tools to find out who's doing what.

Software development is about more than just writing code. It involves thinking, solving problems, creating systems and algorithms, providing solutions, and then coding them. So, whatever analysis document you have, you'll have to go through this process every time you face a software development challenge.

This means you may not know what a development team is doing just by looking at their screens. It's best to talk to them on a regular basis to understand what they're doing. The agile development methodology and scrum meetings can be used to achieve this goal, even if you're working remotely.

After the global pandemic struck in 2020, most companies moved to remote operations and many of them won't be coming back to offices even after the pandemic ends. So, you'll need to have a remote working system in place, even if you have an office.

Using tools to track what people are doing may also create an ethical problem because everyone needs privacy, even at work. We have to find a way to keep the process in control while respecting people's privacy.

What I use is a mixed process. I require everyone to show me what they did in the course of a day, but I don't track them. I suggest they track themselves, however they want, and then show me the results. With proof, of course.

In a development team, having the proof is relatively easy. You just take a look at the commits and ask about the intervals. There may be some working out to do in allowing for small commits or seemingly big commits at a fast pace. You can understand what's going on by communicating with your team on the commits. This way, everyone will be on the same page.

Understanding things like time spent and work done is a profession, and it's one which, if needed, you can learn. The name of this profession is "management" and mastering it helps everyone have a clear understanding of the processes they are involved in.

Now, let's discuss the testing and acceptance processes in a development workflow. After that, we'll finish the development phase and start the post-development phase. And after that, I'll give you tips on automating your workflow so that you can do numerous things which don't involve too much work, and can maybe be done just by yourself.

Testing & Acceptance

Testing and acceptance are two integral parts of Agile methodology. They both have different types and workflows, and they're fun because the subject of testing and acceptance is the work that's been done. Let's dive in.

Testing

Testing can have different forms. Developers start with developer testing first. It's the testing process done while developing. It's like writing code and running that code to see how it works. It's also called manual developer testing, since it can also be automated.

The automated tests are like unit tests, integration tests, and so on. Unit tests are one of the most debated parts of software development. We'll take a look at it first.

Unit Tests

Unit testing means testing the minimum amount of functionality as possible. It's like testing every line of code you write. Every developer knows how vital unit testing is, but most don't write tests.

There are some debates about where unit testing should be in the project. There is an approach called Test-Driven Development (TDD), in which you write tests of what you want to code before you write the actual code. There are rules like "don't write a line of production code without writing tests" and suchlike. It's a hardcore practice, and not every developer can cope with that kind of testing routine.

Another approach is writing tests after writing the code itself. It's easier than using TDD, but may be weaker when it comes to the tests' coverage. However, almost no one completes writing the test suites if they're written after the code.

Let me explain these approaches individually:

In Test-Driven Development, you start by writing a small test and then you might see it's not "passing." Passing a test means it runs well without any errors. When a test passes, it is green; it's red when it doesn't. After writing a small test and seeing it has failed, you need to write the minimum amount of code needed to pass that test.

It may not have the functionality you need, but it's just the minimum amount of code you need to make the test green. Then, you write another failing test and make it pass with the actual code.

After you have passed enough tests to form a functionality, you start to "refactor." Refactoring means writing the same functionality, but cleaner and more understandable. After the refactor, you test again and see there are still all greens, and you move on.

TDD is a slow practice, but if done well it's a great one because it gives you the ability to release the code with confidence because of the tests you have written.

Another approach, the regular testing practice, is the quite the opposite. You don't drive your code from the tests; you just check if everything continues to work with testing. You may choose to write tests for the essential parts of the app you have developed, or you could write tests for everything; the choice is yours.

Integration Tests

Unit tests fall short in some real-life cases, for example, if you want to test the user interface, it can't be done by writing unit tests. That's when integration tests come to the rescue.

Integration tests are the things you write to test the product from the user's perspective. It's like touching the buttons, filling in the forms, shaking the devices, and so on, all with code. It's more fun to write and to operate than unit tests, but it's not utilized by many developers.

You can write integration tests for almost all platforms. You can open up a test browser on the web and automate the flow of actions to follow. The test suite will open it and do whatever you want it to do. You can also see the actions by yourself or make them "headless" (which means not in plain sight).

In iOS and Android, the test suites open up the device simulators (or real devices), install the app you have developed, and do whatever you tell them to do.

Testing overall is a good practice in programming. If you test your code, you make sure everything is working. Even when things get complicated, you can trust your tests and modify the code as needed.

However, writing useful tests needs practice, and as it's not viewed as being as valuable as coding itself, developers can easily underrate it. The primary rule here is to prefer someone who writes tests but doesn't require them because it may dramatically decrease the number of developers you're addressing.

Acceptance

Acceptance is another practice in Agile, along with testing. It means you test the functionality by hand and literally "accept" it so that the process can move on.

In the Waterfall development methodology, acceptance tests are put in place after developing the whole project, so the process becomes as hard as the development itself. In contrast, Agile methodology places acceptance testing after the sprints.

* * *

After every sprint, you look at the product, test the functionality within the sprint's boundaries, and accept or reject them. If you accept all of them, the sprint will end. If you reject some of them, the project manager will need to decide what to do.

There are two things they can do after you reject some of the tasks: They either decide to solve the problems immediately, or they decide to put the desired functionality to the backlog as high-priority tasks. Then in the normal sprint planning process, they can select the tasks and move on.

In well-planned sprints, a timespan is reserved for testing and solving bugs that appear inside the sprint. So, it won't be a problem to reject a task if you're working with a good project manager.

After accepting the whole project one task at a time, it's time for its release. Releasing an app or a website to the public is not an easy task, and you may want to release it to small groups as alpha and beta releases before the actual release itself.

After all, users' acceptance is the most important thing for a product.

Let's have a recap of what we've discussed in this chapter so far.

Chapter Recap

The development phase is something you'll face again and again in your entrepreneurship journey, probably with multiple projects, so understanding what's going on is essential. Here's what we've discussed in this chapter:

1- Time Estimates

You can live without time estimates, but living with them is a more comfortable place to be. That's because, as an entrepreneur, you'll need to have an idea when the development process is going to produce a product and when you can launch it.

* * *

Having the right time estimates every time is tricky, so it shouldn't be the strict criteria. Rather, the criteria should be persistence when it comes to making time estimates. You can do things to help your team ease the time estimating process, and we've listed them all in this section.

2- Talking the Same Language With Developers

Development isn't black magic; it's just telling the computers what to do. It's all about trial and error, having bugs, and fixing them. So, developers are problem solvers, like you are as an entrepreneur.

If you listen to them and do your homework about your areas of responsibility, you and your team will eventually be talking the same language.

3- Managing the Project Together

You shouldn't manage your team; you should manage the project with your team. The main difference here is the same one which distinguishes a leader from a boss. It's essential, because as a leader you can motivate people to make your dreams a reality.

It would be best to be consistent in your actions with transparency and responsibility. If you know everyone's responsibilities and perform well, it'll bring the whole team to a level where they'll work like you. Giving and taking honest feedback can help communicate what's going on.

4- Where to Store Codes

Storing codes is an easy task, but you do need to know what you're doing. You can store your entire codebase in cloud services or your servers. With version control systems, storing code securely is not an insurmountable problem. I suggest you select a provider and store all your codebase in it.

* * *

5- Usage Rights of the Codes

In short, it's a no-brainer that the usage rights should be yours from the start, but the problem is not about the rights, it's about the ability to use those rights.

If you don't have the code in the cloud, you can't claim ownership of it. If your team say that they'll push the code to the cloud after they finish, it means you won't own the code before the project finishes. Be careful about those situations and try your best to get the codes immediately after someone writes them.

6- Project Management Techniques Overview

There are innumerable project management techniques out there, but we've discussed only two of them: Waterfall and Agile. It's because they're the most popular methodologies among development teams. They have their advantages and disadvantages, but these days, for a project that is just "complicated enough," I suggest you utilize Agile methodology.

7- Agile Project Management & Scrum Methodology

Agile originated at the beginning of this century, so it's no longer new. It gives you cycles, which are called sprints, to manage your work in parts. Dividing the work into task groups and having a continuous release cycle makes it so much easier to adapt the project's timeline requirements to changing circumstances.

Using Agile methodology can be challenging with small teams, so you might care to utilize "kanban" as your go-to project management methodology. It involves having lists of tasks and moving between them continuously.

8- Sustainable Development

Even if you are the project's only programmer, it's still necessary to write good code. Good code means it's extendable, sustainable, and understandable by people without prior knowledge.

* * *

Sustainable development means that you can still develop your project even if the conditions change. The development team may change over time but if you have a good codebase it won't affect you much. However, if you don't have a good codebase, you'll have a hard time hiring new people to work on it.

9- Code Reviews

Utilizing code reviews is a great way to make sure everyone is on the same page. If you have a development team instead of a single developer, making code reviews is easy. If you don't, you can utilize selective reviews to review the code yourself.

Putting comments in code is a great way to communicate with people via the code you write. Utilizing comments in the code is also a great way to ease the code review process.

10- Quality Assurance

Assuring software quality is an ongoing task. Coding by yourself for yourself is a recipe for creating garbage software, so you need to have a code review cycle of some sort.

If you're not working with professionals in the review process, you can perform random checks to ensure everything's running well and as expected.

11- Work & Time Tracking

There are many ways to track who's working on what, but you also need to protect people's privacy within your organization. So, the tracking should be done voluntarily, as an integral part of your organization's culture.

Work tracking is you making sure everything is proceeding as expected. Time tracking is you making sure everyone is working as expected. There will be times when you need both of them.

* * *

12- Testing & Acceptance

Testing can have different shapes and forms. You can utilize automated testing for your projects and it's a great way to release your software with confidence. You can also have manual test cycles at the end of each sprint, these are also called acceptance.

Acceptance means you test the work done and accept it as it is. Make sure you know what to test and that you have tested everything you need to test before accepting it.

Now we're moving on to the post-development phase. We'll take a look at the documentation, working with new developers, maintenance agreements, publishing and hosting your app, a thing called dev-ops, and so on. Let's dive in.

CHAPTER FIVE

After Development

Documentation

In 2018, an entrepreneur friend called and told me that she needed help. When I arrived, she gave me some more information about her situation. She was an experienced tech entrepreneur and she was working with a brilliant development team.

Her team was a skilled one which had been working together for more than a year, and everything seemed to be going great. One day, however, the head of the development team got an offer he couldn't refuse and accepted it. He also took his team with him, and as a result my friend was alone and left with the code they had been writing for more than a year.

She tried to hire different people to continue with her project, but without success. So, she called on me to explore why no one wanted to work with her codebase When I looked at the code, I immediately realized it was bigger than I first thought.

The development team had worked hard and they'd done an excellent job of making a great product together. The process they used was good; the code quality was sufficient, they used code reviews and all the necessary processes needed to make a great software product. Except for one thing: documentation.

There were countless lines of code but no one had written a single line of documentation about them. The code was excellent, and you could understand the things they were doing on every line, but there was nothing written down about the project at all.

I asked my friend about the process they had followed. She told me that first she hired the lead developer as CTO, and then he brought in his team of developers. They talked their way through the product and the developers started immediately after the first meeting. They iterated, iterated, and iterated. They created a massive codebase quickly, and everything was fine. Until they left.

Then I had to hire some developers to work closely with us to document every feature in the code, which cost us more than a month's worth of full-time work. It turned out there were some unnecessary parts in the code. They had a lot of duplications, and unused features that no one had the courage to remove from the codebase.

Everything became clear at the conclusion of the documentation phase. We were tired, but it was worth it. After completing the documentation phase, we formed a new development team, and they started to work after just a week of the onboarding process.

Software projects get complicated with time, at least that's the general rule. If you have a project that you don't want to extend or develop further, you could skip the documentation process because it would be worthless for you. However, if you plan to develop your product further, solve its bugs as soon as they appear, and expand your team in the future, you need documentation.

Documentation can be originated from the code itself, or it can be a completely different document; the choice is yours. If you want to document every function you have, there are ways to extract documents from code comments. You just need to pick a system and follow it. After you're done, you'll have documentation written with code comments.

Nowadays, no product is confined to its own small environment, almost all of them are somehow connected to another. You may use a backend service or utilize login with social networks. These are all integrations. It's a big ask to produce something that's completely isolated these days.

You'll eventually need to update the versions of the frameworks, libraries, and Software Development Kits (SDKs) from the services you use, so you'll eventually need to update the code you have. Documentation helps you fill   new requirements easily when you update the third-party libraries.

You don't have to be the one who writes the documentation itself, but as an entrepreneur you'll need to learn what's going on there. Here are some quick tips on how to make great code documentation:

1- Start With the Code Itself

Product documentation starts with the analysis you prepared before the development began, and code documentation starts with the first line of code. If your fellow developers know that they'll eventually need to write documentation, they'll write explicit comments and human-understandable code to fulfill that requirement.

It's essential to have the developers on board because they write the code and they know what exactly the code does. However, they'll eventually forget. I have written so many apps, websites, machine learning and deep learning applications, small programs and big programs, that there's no way for me to remember the reasons why I wrote each line.

That's the same with your fellow developers too. Even in your project, they may write an amount of code so large that they forget why they put something in there. So, it's best to write the reason down after writing the code. It eventually makes for a smooth route through the documentation if done right.

2- Use Documentation for Refactoring

One more thing the documentation is good for is refactoring. We've discussed refactoring in previous chapters; it's adjusting the code you write to make it clearer, more understandable, and more extendable.

* * *

You can utilize the documentation-writing process to check the code and refactor it the way you want. Refactoring can be a great way to spot unnecessary code and you can write it into the documentation after fixing it.

3- Don't Repeat Yourself

It's one of the core rules in development: Don't Repeat Yourself (DRY). Computers don't care if you repeat a twenty-line block a thousand times or not, but it's painful for human beings. That's why we use functions; to group functionality to reuse it again and again.

So, you utilize the documentation-writing process to remove duplication from the codebase. There are many ways you can repeat yourself, so it's a good idea to check for repetition once in a while with a fresh pair of eyes. The code review process also helps with the DRY principle.

After understanding why the documentation is needed, let's move on to the next segment to discuss finding new developers for your project.

New Developers

In 2017, I received a message on LinkedIn from a guy who was trying to find some new developers for his ongoing project. He described his project and the situation he was in, and I was impressed with how organized and well-prepared he was. He wanted me to form a team of developers.

I began by analyzing the situation. Previously, he had three developers, and, one at a time, they dumped each other's code and started to rewrite the project from the ground up. So, we had a project that, despite being written three times, was still incomplete.

He messaged me in the first place because he couldn't build a system while multiple people were trying to improve it. We needed to create a system for him, so new developers wouldn't decide to dump the old code and write everything from the ground up.

We ended up in a good spot. Let me tell you how we did it by sharing some tips about the process:

1- Have Analysis & Documentation at Hand

It's essential to hire new people for your startup in the tech industry. If you have a well-documented project, it'll be easier to find out whether someone is interested in it or not.

Finding someone who is interested is vital because we want them to improve on the codebase, not rewrite it. The project may not be a big one, and rewrites will be necessary from time to time, however, we don't want to simply dump the time spent on the project that we don't have to, because time is the most valuable thing of all.

After you have worked on your analysis document with your fellow developers, work on the code's documentation. If you want to make sure everything is going well with the documentation, you need to read it. I know it seems onerous, but it's the way you can ensure something in a small business is correct, particularly if there's no one you can trust to do it for you.

Analysis documents and product documentation make it easy to find new developers for existing projects, and they also make working with them a great experience. That's because developers shouldn't have to think about business decisions while coding. They just need to code, and while they're doing their jobs, you'll be responsible for all the business decisions.

2- Look for Hard Work as Much as Talent

I learned from life as a talented guy that talent is cheap. Yes, it can get you somewhere fast, but only if you're willing to walk the walk. It's the same when you need to find new developers for your existing projects.

* * *

Released projects are easier to maintain, and we'll discuss that in the next passage. Developing a half-baked project is a challenging task which is not for the faint-hearted, so you need to find someone with the ability to work hard.

Working hard is a thing that enables a person to do whatever they want, but it doesn't happen overnight. You can see the difference between a hardworking person and others when you interview them.

Hardworking people give you the ability to focus on your responsibilities while they get on with theirs. That's the key to success for a new tech company: Everyone has to do their job well first. You can and will do others' jobs, but that doesn't mean you should neglect your own skills.

However, knowing your craft doesn't grant you extended deadlines, and you may need to act quickly from time to time. The same thing happens to developers, so you need someone who works well under pressure.

Hardworking people cope well with working under pressure because they care about making their work into something great. That's the attitude you want to see when looking for new developers for an existing project.

3- Set Goals

You may make some wrong decisions when hiring new developers. You might choose to work with the wrong people, maybe repeatedly. So, it would be best if you had a framework for you to work with which would tell you when to get rid of them.

If you set goals and agree them with the people you work with, you can check their progress and act when it becomes clear that the agreed goals are not going to be achieved.

It's a smart way to find someone willing to work on your project; however, it has some caveats:

* * *

First of all, you should set achievable goals because the people you'll be facing are smart. If you have unachievable goals, it'll be tough for you to find someone. Second, you need them to agree on the goals that you have set. Otherwise, you won't be able to judge them against those targets.

Setting goals keeps you aligned with your team, and when you're aligned with your team, you'll be there when something's not working. That's the main point of setting goals: You can see the progress towards them, and you have the necessary data to act upon.

4- Track Every Metric Necessary

To get that data which you may need to act upon, you'll need to track every metric necessary to see what your company is facing.

Tracking metrics is different from the tracking work and time processes that we discussed earlier. Your project's metrics are the progress bars you want to fill. They're the numbers you need to increase or decrease to achieve your goals.

If your goal is to create a product within a timeframe, your metric will be the task number, and you want to get it to zero. If your goal is to make the user onboarding process faster, the metric is the time it takes to save information to the server and your test users' happiness about the process.

You can get creative on goal setting, and you can make every bit of the flow accountable. Every accountable thing can be a metric, such as happiness. You can find a way to track your test users' happiness, by asking them to fill a questionnaire, perhaps, or by watching them while they use your product.

If you have goals, you may eventually achieve them. If you track metrics to achieve those goals, your chance of achieving them is much higher.

* * *

Maintenance Phase

While writing good software is a challenging task, maintaining its quality level is even more demanding, and developing it further while maintaining that quality cranks up the challenge even higher. So, how can we find our way around in the tech business while these things are so difficult?

It's a good question. That's why we discussed the work that needs to be done before the development aspect; analyzing the project is as important as what the developers do.

Without adequately analyzing a product, you can't create a maintainable system, at least most of the time. If you want to have a system designed for future development, it's best to act like that from the very beginning.

There are two main strategies when it comes to maintaining software. Handing it over to an entirely new development team or keeping it with the team that developed it. Either one has its pros and cons. Let me give you some tips about each of them.

1- Hiring New Developers to Maintain the Code

Your contract with the development team may terminate when the development phase ends, or simply because you don't like working with them. No matter the reason, there should always be a way for you to hire new developers to do the work.

The ease of finding new developers is directly related to the code's quality. Because, after signing an NDA, the developers will want to see the code they'll be maintaining. If the code quality is low, they may want to be paid more to maintain it.

Maintaining is about solving bugs and adding new features. Adding new features is a bit easier, but solving bugs in messy code is a nightmare. That's because you can't be sure that what you're doing is right. You can solve a bug, but it may create another bug in a completely different part of the software, and no one will notice until it creates another problem.

* * *

Either you do the analysis in the beginning and the documentation in the development phase, or the new developers have to do those when they start. Creating analysis and documentation takes more time and resources if they are done after the development phase.

Because analysis and documentation are also guidelines in developing the software, you're also guiding yourself through the journey if you create them alongside the development phase. You can spot inconsistencies, flaws, and gaps in the software while writing the documentation.

If you have those documents at hand, the process will be more straightforward. If you don't, you'll have to allow the new developers to create at least the documentation. The analysis may not be necessary for small projects, but documentation is a must for developing with more than a one-person team.

You can hire a development company to take the risk, but they may not be with you throughout the journey, so you may want to form your team in-house to maintain the software.

You may want to add features, even after the designated end of the development phase. Adding new features is a considerably more comfortable exercise than fixing bugs, but here comes a caveat: You may be creating more bugs than you expect.

Understanding the code before you develop anything is essential. It'd be best to make sure everyone understands the codebase as if it were their own before developing new features.

2- Working With the Same Developers After Development
Working with the same team after the development phase is a bit easier than working with a different team. That's because of their familiarity with the code they have written.

The level of familiarity will be about the time they spend on the project, but in reverse. The more time they spend on the project, the less they're familiar with it. That's called estrangement. After a certain period of doing something, you lose familiarity with the work you're doing.

You may know everything within the project, or you trust yourself to. Either way, it's dangerous to maintain software without proper analysis and documentation in place. That's because when the project grows, the development team will still be focusing on writing the next line of code. If you don't have a guide that oversees the process the team will eventually get lost in the software they are writing.

You can do two things here: First, you can make sure everything is well-documented. Second, you can instigate a refactor/rewrite phase as a step forward.

The second one doesn't seem like a great idea, but it's one you may consider, no matter what. If you have a small project with a short deadline and putting it on the market was the priority, then the code may lack quality. That's normal because it's about the tradeoff between quality and quantity.

On these occasions, you can write the software again after the first development phase. It's not like writing everything from the ground up, rather it's just copying and pasting code from the old project and making sure of the quality.

Making slight improvements to the software every time you work on the project is essential for the success of the development phase. You may have bugs, or you may want to add new features; it's all the same. Simply improve the codebase every day, even if it occasionally means writing everything all over again.

Let's face it, bugs are unavoidable and you will have a bunch of them. That's not the problem. The problem starts when you stop developing your product. There will always be things to do to improve the codebase; whether writing tests, writing documentation, refactoring, rewriting, etc. If you stop doing that, it means you have abandoned your project.

Abandoning a project may not be the best thing for you to do as an entrepreneur, even if you're starting a new one. That's why you should form your development team for your projects; because software development is not a one-shot thing, it's an ongoing process. The more you give in the beginning, the more you have in the process. There's no end if you don't quit.

Let's take a look at how you can make your project live, even if it's not complete.

How to Publish & Host Your App

After developing your application, you'll need to figure out ways to distribute it, to attract users, and eventually make money from it. The distribution process is different on each platform, so I'll explain what the process looks like for iOS, Android, and web platforms. Also, we'll discuss the distribution process of the backend system.

iOS Distribution Process

For developing iOS apps, you need an iOS developer account. You can get one by creating an apple id and going to the developer.apple.com website. Then you'll need to choose between setting up as a company or as an individual. The main difference is the name on the App Store and the availability of certain app types, such as simulated gambling.

If you want to have an individual developer account in your name, you'll need a credit card with your name on it. After you enter your personal information, you'll pay $99 upfront and it's payable annually at roughly at the same time. Then you'll have an individual App Store account. You can use it to upload apps and monetize them.

If you want to have a business account on the App Store, you'll need a D.U.N.S. number. It's a number used to identify your business and getting one can take up to two to three weeks. So, if you plan to open a business account before you release your app, it's best to do it sooner rather than later. The business account will also cost $99 per year.

* * *

After you have opened your Apple Developer account, you'll need distribution certificates. If you have a team of developers, you should add them to your developer account so they can use the app signing features of the account.

App signing is a process that ensures you're the developer of the app. This means you'll create a certificate, use it to sign your app, and Apple will know the app is from you when you send it to the App Store.

You'll need to have a Mac for developing any application for any Apple store. That includes iOS, macOS, tvOS, and watchOS. The development and distribution processes are pretty much the same for all the platforms, so you can adapt your processes according to the specific needs of the platform you're developing for.

If you sell virtual goods or subscriptions in your app, you'll need to add them to your App Store app page. You'll need to set prices beforehand and send a screenshot for each of them for Apple to review. You can test the purchases before the review process, so there are no worries about that.

After signing the app, you should upload the app to the App Store via an official program, such as XCode or Application Loader from Apple. XCode is the main program in which you develop your Apple apps. Without it, Apple currently doesn't allow you to distribute to any of their app stores.

After uploading your app to the App Store, you'll need to have a store listing page. You'll have the app name, description, keywords, app screenshots, etc., to complete, and after you've done that you can send your app to the review process. The review process typically takes a day or two to complete. If your app meets the requirements set by Apple's guidelines, they'll approve your app and release it. If you want to release your app manually, make sure to check that option on your app info page before you send it for review.

If they reject your app, they'll send you a message from the "resolution center," in which they'll describe why they rejected your app. Details about any problems are provided, perhaps with screenshots, together with a description of the next steps you need to take to solve them.

After you have solved the problems, you can send the app back for review, and they'll review it again. The review process is the same, even if you want to update your apps.

After you succeed in the review process, you'll have an app on the App Store.

Android Distribution Process

There is more than one store for distributing Android apps. The instructions I am sharing here work for the Google Play Store. You can also use the core ideas here to distribute your app on other Android app stores.

For distributing the Android apps on the Google Play Store, you'll need a developer account there. You can go to play.google.com/apps/publish to open one. It costs $25, which is a one-off, final payment with no annual recurrence. You can use individual or company information because in the Google Play Store it doesn't matter which. If you fill in the information correctly, you can open a developer account in minutes.

After your developer account is ready, you'll need signed "apk" or "aab" files to distribute the apps. They can be created without a certificate at hand, but there's a catch: You'll need to have a "Keystore file" to distribute the app. You can create it from any machine, but if you lose it, you're in trouble. Don't ever lose the Keystore files you used to develop your apps. Create them yourself and store them in the safest place you know. You can also have backups of those files, and I recommend you do. You'll need the Keystore files to send updates.

After sending an apk or aab file to the Play Store, you can add your in-app products from the Google Play Console. The way you do that is pretty much similar to the App Store process, so I won't go into those details again.

* * *

Your in-app products are set, now you'll need to have a "closed testing process" to test your app's purchase flow. Closed testing is simply uploading your app to a closed test track and downloading it from there to test the purchase flow. The details change from time to time, so I won't include them here. You can check out their own documentation for details of testing purchases.

After you test your app, you should complete the store listing page. After that's done, you can release your app to the public.

There are additional steps required on both platforms about the data you use, your target audience, etc. Be prepared to fill some forms here and there.

Web & Backend Distribution

For releasing web apps, you won't need any of these. You just need public servers where you can put your code, and then release your product.

However, if you want to develop more than a simple website, or deploy your backend services and scale them according to the load you have, you'll be entering another field of expertise called "dev-ops."

We'll discuss what dev-ops are and how to find dev-ops people in the upcoming chapters.

What is Dev-Ops?

Dev-ops is a term consisting of two terms: Development and operations. We've already discussed the development. Let's see what's involved in the term "operations."

An operation in the software world means the systems that run the code. On a website, it is running the servers in the background. In a mobile application, it's running the backend server.

* * *

The operations part is as important as the development element because it enables the code to run on different environments to reach its audience. It's a profession on its own, and there are languages, frameworks, and tools to learn if greatness in operations is your goal.

Dev-ops, on the other hand, is a term describing a workflow. It's the distribution side of Agile development. It has five steps: Build, test, release, monitor, and plan. You can think of it as an agile sprint, but in the clouds. To explain:

In 2019, one of my clients had a problem with the servers where they had put their website. They're great at business and marketing, and the website was working fine, but whenever they worked with an influencer on YouTube, the website went down.

The servers are on a great cloud provider, and they're spending a good amount of money on them. But the problem was that sometimes it wasn't enough. Everything was fine on a typical day, but the website often went down on marketing campaign days.

That's a big problem for a startup. Even if you have everything, and even if you can do great marketing campaigns, you have to have a system in place to support you. The thing that was needed for that client was auto-scaling.

Auto-scaling is a method used by cloud infrastructure providers that offers you the correct scaling, whatever your load is. If you have ten people visiting your website, you'll have just one small server and you'll pay for just that one small server. And when the time comes, for example, after a successful marketing campaign, the servers automatically scale up to cope with the load increase.

It doesn't matter if there are a million people on the website, it continues running because you have auto-scaling in place. The best part of auto-scaling is that it can scale down and up, so you won't pay any extra money when you have no visitors.

That was a problem solved with operations. However, there were other issues, as my client wanted to develop the website further. The team of developers knew their stuff, but they didn't have a continuous way of delivering the work they'd done.

Which means it's time for dev-ops. Dev-ops is the glue between development and operations. By using dev-ops when you develop a feature, you test what you are developing by deploying the code to test servers and, after the test, you release it to your customers. You continuously monitor the processes and catch the bugs or errors which are present and make plans to solve them in your next release.

Dev-ops is the thing that enables you to carry out this process with a couple of commands. It's essential, because of something called "environment."

An environment is a set of conditions in the machinery you're currently working with. Let me explain:

For example, you have a development team consisting of three people. You're working with the Python language and the Django framework. One of them has a MacBook Pro, with Python 3.7 installed. The other one has a Windows system, with Python 3.9 installed. The other one is using Linux with Python 3.2.

There is a codebase with some libraries in place. When someone writes a code, they push the code to the cloud code repository for everyone in the team to access and develop further.

However, the person who has a computer running Windows can't run the project. Most of the time, the reason is the versions of the libraries that are being used. The project can be configured with Django 3.1.6; someone can have Django 2.1.15 installed on their computer. This breaks the development process.

To deal with this problem, smart people set to work and invented environments. An environment consists of everything necessary to run the project, from the operating system to the code libraries used. If you use the same environment everywhere, you just exchange the code, and everything will work just fine.

It's not easy in complicated projects, and it's better when you think about it from the beginning. The computers you develop will not always be the same as the computers you deploy.

Dev-ops is the process that utilizes environments to push code from one computer to another. This smooths the way work is done on the project, as well as deploying it to the servers. It also solves problems like the downtime you face while updating.

When you write code on a computer, it runs on those conditions. If you arrange the conditions and make sure they're the same everywhere, the process of moving the code from one computer to another will be as pain-free as it can be. However, not every computer is made equal, and you may not want to run Linux on your Windows machine every time you start developing a project.

This problem is solved by "containers." A container is an isolated environment that has an operating system layer within, so you can run any Linux container on your Windows machine with all the necessary libraries that come pre-installed. The best thing is that containers can be used by every computer, including servers. Meaning, when you want to deploy the code to the server, you transfer the container and everything works as expected.

Let's find out what's used to build such containers and the tools you need to use to have a dev-ops team in place. Dev-ops tools and frameworks are best when utilized by their teams, but your fellow developers may have the ability to use dev-ops methodologies in the process.

The main point here is that you have a continuous deployment and testing process, like in Agile development methodology. With that in place, you can work confidently on your project and know that you're well-placed to act fast should any problematic situation arise.

Now let's look at the languages and tools for dev-ops.

Languages & Tools for Dev-Ops

In dev-ops processes, there are hundreds of tools which you can use to your advantage. Some of them make the containerizing process more straightforward, some of them orchestrate the containers, and some give you reports and alerts.

Before discussing the most popular ones, we'll have a look at two terms: "Continuous Integration" and "Continuous Deployment" (CI & CD).

Continuous integration means making sure every member of the development team has the latest version of the code. It's the process of syncing the code between the team members. Having a version control system is the first step, having a code review system in place is the second. With a continuous integration process, you'll make sure everyone is working on the same project.

Continuous deployment is pushing the project to the end-user. There's one other step called "continuous delivery" between CI & CD, but for simplicity's sake, we won't separate it from continuous deployment. In the CD process, you make sure your project is building automatically, passing the tests, and deploying to the end-users, all automatically.

Making something automatic is a challenging but rewarding process, and we have tools and languages to make it less exacting.

The languages widely used in dev-ops include, but are not limited to: Go, Rust, Scala, Python, C/C++, and Ruby. These are used because their speed and capabilities allow better use of the resources. Not all of them are equal in processing speed, but there's a tradeoff between development speed and processing speed.

There are other languages used in dev-ops processes, but these are the most popular. If you're hiring dev-ops people, you'll want to make sure they know at least some of these languages.

Let's talk about the most essential tools.

Docker

Docker is the most popular container platform out there and it's improving day by day. Docker was written with Go in 2013 and is used widely in the containerization process. Docker allows you to bundle your project in one or more containers and manage them with everything you need to include.

A docker container can have the operating system, the libraries, runtimes, and everything in between to run your project. You can create a docker container and put your configuration options alongside your code, and it will run smoothly.

It also gives you the ability to distribute the containers, which means if you have a machine that runs Linux, you can use the same containers built on a macOS or Windows machine, thus getting rid of the pain of managing every environment you have. You just create an environment and use it wherever you need.

You can use the same container in the development process, testing process, and deployment process. It's easy to create and share containers since they're pretty small on their own. Even if you add numerous things to it, you can create an image of the container and store it somewhere, so, there's no need to transfer big files back and forth.

The place where you put your containers and images is called the "registry." You can have your own registry or use a managed one from a cloud provider, and you're good to go.

Kubernetes

Kubernetes is a container orchestration engine. It manages the tasks like no-downtime updating, distributing the load, scaling the containers, etc. Developed to manage docker containers, it's highly popular in the dev-ops world.

You don't need to integrate everything on your own; the docker to Kubernetes integration is already in place.

Let's take a look at a couple of scenarios where you can use docker and Kubernetes.

* * *

Scenario 1:

Let's say you have a team of four developers and you're working remotely. Everyone has a different environment on their computers, with some using different distributions of Linux, some using macOS, and some using Windows. You also have a server to which you deploy the code for your users, and it's running on Linux.

The main problem here is matching the environments on the operating system level. Because if you don't do that, you can have dependency problems that are hard to solve after the development phase.

So, you use docker. You create a docker container from your project and distribute it everywhere. Everyone in the development team on the same container pushes the code to the repository, they review each other's code, and your CI & CD pipeline automatically deploys the container to the registry and distributes it.

This process gives you peace of mind, and you ensure everything goes smoothly without any conflict in environments.

If you want to attain the same peace of mind in deployment, you should use Kubernetes. Kubernetes allows you to distribute the code to multiple containers, even to multiple parts of the world, if necessary.

Container orchestration is a method that makes the deployment process stable and scalable. You won't have a solo struggle with environments or scaling if you configure your Kubernetes clusters right from the start.

Scenario 2:

You have one fellow developer to work with. To make sure everything in place, you use docker to manage the containerization process and ease of deployment. You use a managed container registry and auto-dev-ops features from your cloud provider to deploy the code to production servers.

* * *

This gives you ease of use and an increase in productivity if your team is small. Because when you work on deployment, you can't work on development at the same time. Focusing on development with a solid deployment system in place is an excellent thing for the project.

As we discussed above, the dev-ops process is the tooling and automation process you need to set for your project right from the start. However, there may be times that you don't need containers to make everything run smoothly.

You'll need continuous integration every time you're working with the software. Continuous deployment is optional, but it's nice to have. You don't always need to utilize containers or container orchestration.

Let's take a look at when you need containers and when you don't.

How to Determine the Need for Dev-Ops

As an entrepreneur, you need to think hard about every penny you spend. If you don't really need dev-ops then you shouldn't have it in the first place, so it's important to determine the need to utilize dev-ops.

There are some conditions where you wouldn't need dev-ops. Let me give you an example.

One of my clients wanted me to audit their cost structure about the app they were building. When I took a look at the project, I realized the reason why they called me immediately. They were spending hundreds of dollars on a non-existent app.

If you're working on artificial intelligence, using server power before releasing the app is understandable. However, if you're just trying to have an app with a backend server, that's too expensive.

They were doing everything right with the dev-ops process, which was the problem. They had load balancers, on-demand servers, serverless functions, a Kubernetes cluster, dev-ops tools, and many more things in place, and all in the correct order.

If they had an app that needed complicated work on the backend side, all that spending would be understandable. However, if there's not much going on in your app and you just need a simple solid backend server then even running your servers may be overkill.

I suggested they use Firebase, a great tool from Google. Firebase gives you a stable and scalable backend system, and even if you have millions of users (I tested that personally), it can handle every process like it's the first one.

For a mobile app that utilizes the backend service for syncing data, having files around, and running some not-too-complex operations. Firebase is a perfect fit. It removes the pain of having your own servers.

I suggested my client get rid of their current dev-ops system and move to Firebase. They did, and their costs went down to zero for at least four months that I'm aware of. Only after they reached the limits of free tiers of Firebase products, did they start to pay for their usage.

If you optimize your processes, your costs go down. That's a golden rule. It doesn't matter if you're just getting started or you've been in the market for years, you should be trying to optimize your processes every day. Even a 1% optimization means a lot in the long run. You can take the money you saved and spend it on expanding your product's horizons.

In this case, we looked at the possible overuse of dev-ops processes and how they can harm your business in the long run. Also, dev-ops processes don't only consist of having servers and operating them. It's the "CD" part of the dev-ops process. There is also the "CI" part, which you shouldn't ignore.

CI, meaning Continuous Integration, is an integral part of the software development process. It's not about the servers, it's about people, and the code people have written for other people. Machines understand code even if you mess up, but humans can't, so you need to have a system in place to do the right thing when it comes to managing the code.

In the case above, when they moved their backend systems to Firebase, they needed CI (and even CD) tools to work with Firebase. We started with their ongoing CI process, optimized the code review cycles, and created a pipeline to have better quality code.

We also used Fastlane, an excellent tool for continuous delivery and deployment of mobile applications. We integrated all their products with Fastlane "lanes," and they were good to go.

The process went like this: Someone gets the latest copy of their code, they develop what's necessary and commit the code with an understandable commit message. They push the code back to the cloud repository, utilizing a branching model named "git flow," Git flow is an excellent plugin for Git, and I highly recommend giving it a try.

After they push the code to the repository, someone other than the developer has the task of reviewing the recently pushed code. When that person has reviewed the code, they can approve, disapprove, or discuss. We used Github for all the processes because it's an excellent fit for cases like this.

After approval, the code will be added to the branch for the next release. In mobile development, you can't have "continuous" release workflow because app stores will review all the releases and users will need to download the updates explicitly. Sending updates every day is not the best way. Updating every two weeks works particularly well and many big companies do just that.

Of course, in the event of a critical bug you'd need to update the app as soon as possible, but in this case I'm talking about regular releases. Without having "continuous" releases, the CI-CD pipeline for mobile apps has a slight twist: You continuously release the app to the testers, not to the public.

Sending the app to testers is also challenging for mobile apps. Bundling and distributing apps is a time-consuming job and the app stores want to review the apps you release to beta testers. That gives you unnecessary review times and it slows you down.

Luckily, Firebase has a tool called "App Distribution." With the app distribution tool, you simply upload your bundle and distribute it to whoever you want. It needs a bit of configuration initially, but after that, you can integrate it with your CI-CD pipeline and run everything from one place. You can deploy new versions to testers at intervals of your choice, even if you want to do it daily.

You don't always need to have dev-ops processes in place, but it's ideal to have the right mindset on how things work for your specific needs. If you're developing for the web, you may need a CI-CD pipeline to streamline your releases. If you're making mobile apps, you may need it to distribute your apps to testers.

It all comes down to the point where your unique needs come into play. Your unique needs should govern the process of creating and managing pipelines. If you do the setup right, you'll be good to go.

Let's have a recap of the "After Development" chapter.

Chapter Recap

After the development phase, you'll face the maintenance phase, which consists of bug fixing and adding new features. You may also eventually need to rewrite the software, but when that time comes around it won't be a problem because it'll be about expanding your horizons. Here's what we've discussed in this chapter.

1- Documentation

Documentation should be an integral part of your software development workflow because there will be fresh eyes to look at your codebase. When new developers come, they have to understand what's going on if they are to continue developing the project.

That's also why you need to understand what's going on in your codebase as an entrepreneur. You'll eventually need to tell new developers what's going on, and documentation makes that process more straightforward.

Even if you're planning long-term collaboration with the same team, you still need to have documentation. That's because developers can forget about code that they have written. That's normal if you focus on the task at hand.

2- New Developers

Almost nobody wants to develop on someone else's code, so finding new developers for your project is challenging. You need to have a comprehensive analysis document and documentation at hand to make sure you can make them understand the existing codebase.

You should look for a hard work ethic as much as talent if you are to find someone who will fit into your project. Finding hardworking people is more complicated than finding talented people, so remember that talent is cheap.

You must also set clearly defined goals to work with new developers and track every metric necessary in their work to make sure they fit with the project.

3- Maintenance Phase

The maintenance phase consists of solving bugs and adding features. It's harder to maintain the software's quality than it is to write quality code in the first place, so you should be prepared for the maintenance phase.

Especially on mobile apps, where it's even harder to send new updates, having a way to solve bugs when they appear and adding features when needed are essential tasks.

You can either hire new developers for the maintenance phase or work with the ones you already have in your team. Both ways work if you do your homework well. Documenting the codebase and having a comprehensive analysis document will help you along the way.

* * *

4- *How to Publish & Host Your App*

In this part, we discussed how to distribute your software applications to the public to make money.

For the Apple App Store, you'll need to pay $99 per year as developer account fees, and you need to have a Mac to publish your apps. You'll use a software called XCode to publish your iOS, macOS, watchOS, and tvOS apps, and there will be a review time of a couple of days.

Emerging from the review period with an approved app means your app has been thoroughly reviewed and accepted, so now you can distribute it on the Apple App Store.

For the Google Play Store, you'll need to make a one-off payment of $25 , and that's all. There'll be no need for any other annual payments to keep your developer account active. After opening your developer account, you'll be able to distribute Android apps on the Google Play Store. There will be rules just like Apple has, and they'll also review your apps before you can release them.

After they have reviewed your app, it will be published on the Google Play Store and you can start making money out of it.

For the web, where there are simply no rules except those governing legal matters, you're free to use whatever service you want when bringing your project to the public. However, you'll need a way to manage your processes.

5- *What Is Dev-Ops?*

Dev-ops comes to the rescue. It's about integrating your development with your distribution process. Dev-ops is a process consisting of two steps (you can make it three if you want): Continuous Integration (Continuous Delivery) and Continuous Deployment.

Continuous integration is about syncing the code between contributors. Continuous deployment is about syncing the code between developers and the end-users. Managing the "syncs" is called the dev-ops process.

You can have a simple server or a big Kubernetes cluster to deploy your code; it doesn't matter. If you plan your processes right from the start, you'll be good to go.

There are ways to utilize dev-ops processes in mobile app development. They're about automating the testing workflows and deploying them to the stores with automation. You can also use dev-ops processes for your backend environments.

6- Languages & Tools for Dev-Ops

There are plenty of languages and tools for dev-ops. Almost all general-purpose programming languages can theoretically be used in dev-ops, but some languages have the necessary tools and well-supported frameworks. Those include, but are not limited to, Go, Rust, Python, C/C++, Ruby, and Scala. They all have their advantages and disadvantages.

Besides the languages, there are tools like Docker and Kubernetes to manage "containers," or blocks of code that contain everything for software to run from the operating system to the frameworks and your code itself.

Docker is the container engine with which you can create and distribute containers to make it easier to have the same environment everywhere. Kubernetes is the container orchestration engine that allows the simple management of multiple containers worldwide from one place.

7- How to Determine the Need for Dev-Ops

You don't have to have a dev-ops process ready for every project. For example, if you have a mobile app or a website, you can utilize a managed backend system like Firebase from Google. There are plenty of tools and systems for you to create apps with ease, even for web applications.

* * *

You'll need continuous integration anyway, because it's about syncing the code between the developers, but you won't always need to manage the distribution process for each part of your system. There are tools to do that for you.

Let's move on to the final chapter, automation. We'll look at automating your workflow and reducing your costs by doing so.

CHAPTER SIX

Automation

What is Automation and Why do I Need That?

After you achieve a certain point in your project, you can automate things to make it easier to manage the process. Automation has several benefits: First, machines don't get tired. They work 24/7 without complaining.

Machines are also cheap and easy to manage. You can set them to work and forget about them, and if you configure them properly, they'll continue to work.

I'll tell you about some automation scenarios to give you an idea about what you can do with machines.

Let's start with something fun. I programmed a bot that makes my room's lights blink when a user spends something on one of my apps. Of course, after a while my apps got bigger and I had to close it because the lights were blinking all day long. That was a good experiment.

I used a service called IFTTT, which was free at that time, to get notified about the payments. Then I made it write every payment to a Google Sheet, add them to Slack, and after that, blink my Philips Hue lights.

If you're making a mobile app with in-app purchases or

subscriptions in it, you'll need a way to verify your users' purchases. I utilized the process to inform third-party systems about a successful purchase. A little coding was needed at the start, but everything else ran without me writing any code after calling a simple webhook.

Let me give you an example from the other side of the spectrum. I coded a video maker for my own use. It does everything you need to create a video by merging video clips, and even posting to YouTube. Let's look more closely at the process:

I started with a video downloader. You put in the video URLs and it downloads the video to your desired location. After downloading the video, I coded a video processor. The video processor changes the video's codec, extracts its audio, and levels it.

After processing all the downloaded videos, I made a video merger. It was the most challenging part, but also the most fun. The merger puts cool transitions between videos. I let the machine choose the transition randomly, but it can be done with parameters. I'm not talking about cut transitions, it's much more complex than that, so making it took a bit of effort.

After merging the videos, I leveled the audios and merged them too. Then, I added intro and outro videos because every great video has an intro and an outro, right?

After adding intro and outro videos, I wrote a music processor. There are plenty of places where you can find royalty-free music, and you can use them in your videos by having a copyright notice in the description; the music processor does exactly that. It finds music, adjusts the volumes to match each other, merges them, and creates a description text including the copyright notices.

After merging the music, I wrote a completer which merges the video and audio tracks. Finally, it adjusts the volume for one last time.

After getting the video ready, you'll need a thumbnail to publish it, right? I made the thumbnail maker for that exact purpose. The thumbnail maker gets a frame from a random point of the video, adds some overlay, puts some graphics onto it, and writes the video's title with a fancy font. Here we go, we're ready to publish the video.

I made this for fun, but if you want to make videos to send to YouTube, you'll need to take two extra steps. You can upload videos to YouTube with their Application Programming Interface (API), but you can't make the videos public if you upload them with an unreviewed app.

If you want to submit your app to YouTube's review process, that may be the best time. But if not, you can upload the videos to cloud storage, like Dropbox, Google Drive, etc., and let an approved application take it from there and send it to YouTube. Zapier is an excellent choice for this exact task. It can watch the storage provider, and when you upload something to somewhere, it can upload it to YouTube.

After the uploading process, you'll need to fill in the necessary information like adding the video to a playlist, scheduling it, adding a title and a description, etc. You can also do that with YouTube's API, and they allow you to do it for public videos. At least they do at the time of writing.

I know, it's hard. It's even more challenging for a non-coder. However, I wanted to show you where automation can take you and I think it's an excellent example for that purpose.

Let's look at a couple of simpler automation processes that you can utilize with your product.

If you have a website and take payments, you should record every transaction and act on them. You can write every transaction to a Google Sheet, post them to a Slack channel (even if it's a private channel), create invoices, and send them to your customers - while also sending them to your archives. After that, you can add a reminder to yourself to personally thank your customer. After all, after utilizing that automation you'll have more time to take care of your customers, right?

If you have an app, you can keep track of your in-app purchases after verifying them. You can create datasets consisting of the purchase and subscription data, and you can create a Google Data Studio dashboard with the data you have. This means you can show your investors the relevant metrics from one simple screen.

You can automate almost everything. It'll be tough to start with, but when you get used to it, it'll be the best thing you can do for your business because it means you'll have more time for other tasks. Having more time is the most essential thing in the world because, for business, relationships, and life in general, the only real metric is time. You can't measure money, success, love, or anything else without time.

How to Automate Step by Step

You can automate many tasks by doing a little work upfront. It's a great way to save money in the long run and you can also generate revenue from it.

Let me give you some details about the path I follow, more or less, throughout all of my automation tasks We'll take a look at how we can find what to automate, how to break it down into smaller parts, and how to automate them. Let's begin.

1- Find Repetitive Work

Finding repetitive work is essential for making a bot. Without using advanced technologies like artificial intelligence, you shouldn't expect a machine to think on behalf of you. You need to have a clearly defined repetitive task to start with.

If the task isn't repetitive, it may not be worth automating it, but it's your call. You'll need somewhere to begin automating, so, to begin with, finding repetitive work is the easiest task.

By repetitive work, I mean the things that regularly occur in our lives and businesses. It may be the invoicing process after a payment; or posting to social media every time you have news from some source. The frequency of the repetition doesn't need to be constant; it can change.

* * *

Further examples could be creating a new user account, receiving an email, putting a meeting on the calendar, or wanting to create videos regularly.

I wanted to create videos regularly with automation, and I'll show you clearly how I approached the task. The first part is knowing what you want, and I wanted to have a system that works on its own and creates videos for me.

2- Break Down the Task Into Parts, Automate, & Merge

Making videos from scratch is not what I wanted; I wanted to merge videos with cool transitions and music. Let's break this down into the smallest logical part.

The first part involves finding videos to download. For that, I chose to use Google Sheets and put the videos on there manually with the parts I wanted to cut. I created a template with a video link, start time, and end time. If you have a trusted video source, putting the rows on the sheet can also be automated.

The second part is downloading the videos to merge, and for this I needed to write a video downloader. A video downloader will get the URLs of the videos, download them, rename them according to a pattern, and make them ready for processing.

The third part is processing the downloaded videos. Not all videos have the same frame rate, level of volume, or audio/video codec, so I needed to process the videos, extract the audio from them, and level the volumes. By doing this, every video I have will be in the same codec and at the same volume level. I also used the start time and end time columns from the sheet and cut the videos to extract the desired parts.

The fourth part is merging the videos. Now that all the videos were in place, I needed to find a way to merge them with cool transitions. The transitions I wanted are complex, so I needed to find a way to apply them. I rebuilt the famous "FFmpeg" library with complex transitions. Then I needed to find a way to merge the audios since the transitions make them shorter.

The fifth part is finding music. There are many royalty-free music libraries out there for you to use with a simple copyright notice. I found 200 different songs that are more or less alike and downloaded them with their copyright notices. Naming them properly was also important since I'd be using them as inputs.

The sixth part is processing the music. I had to make sure all the music files I downloaded had the same maximum volumes, so I adjusted the audio files' volumes while standardizing their codecs. Codecs are essential if you want to merge two media programmatically. I also noted the lengths of the music files for use later on.

The seventh part is adding the intro and outro. These were the videos which I already had, so I added them to the necessary places.

The eighth part is mixing the music. I had the video and knew how long it was, so I needed to select the music files to get the desired length. Then, adding crossfade between them was enough to create a mix. After mixing, I trimmed the music to match the mix's end point with the video.

The ninth part is merging the video with the audio. I merged the two audio tracks with the video after preparing all three exclusively. It was the easiest part.

The tenth part is preparing the video description with copyright notices. I needed to merge the songs' copyright notices and put them into a text file.

The eleventh part is creating the thumbnail image. I had already designed what the thumbnails should look like, and I used "imagemagick" to convert my designs into code. It's not actually converting; I wrote them from the ground up as a shell script.

The twelfth part is cleaning up the workspace. I wanted the system to zip everything for future use and then delete it all except the zips, final video, thumbnail, and copyright notices.

The thirteenth part is uploading everything. I used Google Drive to upload my videos and other files, but you can use any cloud storage provider.

After all the work was done, I merged everything and now I had automation for future use. Whenever I want to use it, I just set the videos I want to merge, and it does the rest.

Finding ways to improve your automation is a challenging yet rewarding thing. I enjoy building automation, so ping me if you think I can help you.

Let's continue with some technical information about the automation workflow. I'll share details of the languages and tools I used to build such complex automation, so you can look at how things work in the background.

What do I Need to Learn for Automating my Workflow?

Automating a video production process is no different from automating any other business procedure. That's why I'm using such a complex example so that you can achieve great things with less than half the work described here.

My automation process starts with the planning phase and breaking down the main task into smaller tasks. Then I start to automate the individual tasks, and finally, I merge them all.

I may be making it look easy, and I know that may not be the case for you, but you can always benefit, as an entrepreneur, from learning something about automation. So, don't view it as being too complicated, just define the required steps and learn each of them one by one.

By automating your workflows, you earn both time and money

simultaneously. So, it's the best way to grow your business on an exponential scale without adding to the workload.

Let me tell you what you need to learn which will enable you to build complex automation like video processing, and you can adapt the process to your workflow. If it starts giving you a hard time, you can ping me and check if I'm available to help you. Hiring someone to help you with automation may produce great results for your business, as I have seen myself with previous clients.

Let's take a look at the tools and languages you can learn and start to automate your workflows:

1- Basics of Programming

Programming may seem like a challenging task, but at the core it's not that hard. When things get complex, everyone, including programmers, has to make it simpler to understand. This means you don't have to worry about not understanding a complex algorithm right from the start.

You can start with the basics by getting any programming book or video tutorial. Thinking like a programmer may take time, but you have to start somewhere. Even writing little scripts will make you feel empowered in front of computers, and that's a great feeling.

You don't need to "start programming" to "learn how to program." You don't have to go all the way and give up all the other things you do to create simple automation or even one more complex. You just need to think about the process like a machine that doesn't know anything about anything.

Machines just do whatever you tell them to do. However, you'll need to give them every little detail to make them work the way you want. That's a problem you'll need to solve if you want to learn to program. You'll need to break down every task into its simplest form if machines are going to do them successfully.

You can start programming with one of the most straightforward and elegant programming languages: Python.

2- Python

Python is an excellent language for beginners to learn how to program. It offers many tools for you to create scripts and run them, and it's easy to learn and understand; it's also competent. You can create entire automation flows just by using Python, and they can work brilliantly for years.

I use Python to create simple automation that doesn't need too high an execution speed. Tasks like automating the home, analyzing purchase data, or creating automated reports can be straightforward with Python.

When you learn Python, you'll see the difference between automating a task and doing it manually. Then you'll get to the point where you see patterns in your daily workflow that repeat. Then you'll easily automate those workflows with Python.

However, there may be tasks where you need a high execution speed, and you may want to utilize a faster language. I prefer Go for automating high-speed tasks.

3- Go

If you want to process and analyze vast amounts of data, for example, stock market data, in real time, you'll need to learn a language which is capable of the highest speeds possible with little work. I prefer Go over lower-level languages like C or C++ because it's faster to write and compile.

Learning Go is easy if you know the programming concepts, but you'll have a hard time otherwise. If you want to learn Go for automation, I suggest you start by learning Python first. If you find Python is not adequate for your needs, start learning Go.

In this way, you can make sure you understand the programming concepts and the differences between them. Go will seem like a complicated language to learn initially, but after you dive in you'll see that it's not the case.

Sometimes, you'll need to use some pre-written programs to create your automation. For example, you may need "FFmpeg," a program written in C, to process, analyze, and edit video and audio files. Or you may need to use "ImageMagick," which is also a program written in C, for processing images.

The best way to use those programs with their interfaces is shell scripting.

4- Shell Scripting

Shell scripting is what you do when you open the terminal or console of your operating system. Every command executes a program written in a particular language to perform tasks.

A shell script is a file consisting of these commands. If you want to execute shell commands in a particular order multiple times, it's best to use a shell script. You simply execute the shell script to do all the tasks.

Shell scripting seems like a challenging task to learn, and it is a tough one to master. However, you can learn the relevant portions of it when you need them.

5- Third-Party Tools

There are also other ways to automate your workflows, and they're widely used in almost every industry. There are tools like IFTTT (if this then that), Zapier, or Integromat which create an automation that consists of multiple steps. I suggest you look at each of them to learn what they do, and I highly recommend that you choose one and use it to automate your workflows.

Conclusion

And, here we are, at the end of the book. Working with developers as an entrepreneur is a broad topic, and diving deep into all the parts will take some time. However, I hope I have given you a good starting point from which to form your development team and make your project a reality.

Let's take a brief look into what we've covered in the book:

In the first chapter, we discussed what to do after you have had the idea. What you do before planning your app is as essential as the planning and the development stages.

After that, we discussed validating your idea without development. Validating your idea may be challenging, but it can teach you so many things about your project.

Then we went on to determining the need for a developer. We discussed thumbnailing and low-fidelity wireframing. Application flows and high-fidelity wireframes came next.

You'll need to have your app designed, so we discussed what you should do to work with a designer and what to expect from the process.

You may need different kinds of developers to make your project a reality, so we took a look at what kind of developers you might need for your project and how many of them you'd need to develop the ideas you have in mind.

After determining the number of developers you need, I gave you some tips for reaching out to them. There may be some developers in your vicinity, or you could find someone to work with anywhere in the world, and that's a great thing about technology.

Finding developers may be an arduous process, so I gave you some tips about how and where to find them, and, once you had done so, what to expect.

In the second chapter, we covered the things you should do before the project. From idea analysis to business analysis to software analysis, you can expect a demanding time when analyzing your project and grading developers.

We discussed programming languages and how to select one. There are no right or wrong programming languages but they all have their pros and cons. So, selecting the right programming language for your project is essential.

We then took a look at the languages and tools for web & mobile development and discussed what full-stack development is and how it can benefit you.

The next chapter was about negotiation. There are different approaches to negotiating with developers, and we covered the most popular ones. Essential parts of this chapter were those on sharing equity and vesting, and you may want to take another look at those passages from time to time to refresh your memory.

Preventing idea theft was the next thing we covered, along with finding more quality development work for less money. Working with consultants on your journey would be a good idea as their expertise can make your life more comfortable.

After the negotiation chapter, we went on to the development process. It's the longest chapter because it is the most detailed one. We started with time estimates, talking the same language as developers, and collaborating with them in managing the project instead of watching over them.

Then we discussed where to store codes and the usage rights of the codes. After that, we went on to project management techniques and Agile project management.

Sustainable development, code reviews, and quality assurance are the topics we covered next, and those are important if you want to have a lasting tech business. In the sections concerning work and time tracking, and testing and acceptance, we covered giving you the control of the process by knowing precisely what's going on.

In the next chapter, we discussed the process after the development phase. Documentation is the key to keeping track of what's being developed, and it's also essential in getting new developers up and running quickly.

Then, onto discussions of the maintenance phase, publishing and hosting your app, and dev-ops. You may or may not need dev-ops, and we covered the reasons in this chapter.

In the final chapter, we took a look at automation. Automation is a fun process, and if you apply it you can generate profits. Automating some tasks makes your life as an entrepreneur a whole lot easier, so I recommend devoting a fair amount of time to it.

If you have a development team in place, they can also help you with the automation process. They can develop the automation and guide you along your automation learning curve.

We've reached the end of the book. I hope it's been a great journey.

I also hope the information I have shared in the book will help you along your own path.

I want to express my huge thanks to those of you who have made it this far. Thank you for your patience in reading and learning. Hopefully, it will all be worth it.

When I started planning it, I realized how much information I had which couldn't be squeezed into just one book. The information on a broad range of topics is necessarily brief because I wanted to give you a firm platform from which to start working with developers as an entrepreneur.

Sometimes it's like developers and entrepreneurs are from different planets but they aren't. As a developer and an entrepreneur myself, I wanted to show you the great ways in which we can work together flawlessly on any project in any scale.

You can reach me at ulascengiz.com, LinkedIn (in/ulsc), Twitter (@ulsc), and Instagram (@ulsc). I'd love to hear some feedback about this book, any new book ideas for me to write, and business ideas where we could work together. I'd love to help you to create great products.

Having to say goodbye has always been hard for me, but we're at the end of the road, for now at least. Until we meet again, stay healthy and stay awesome.